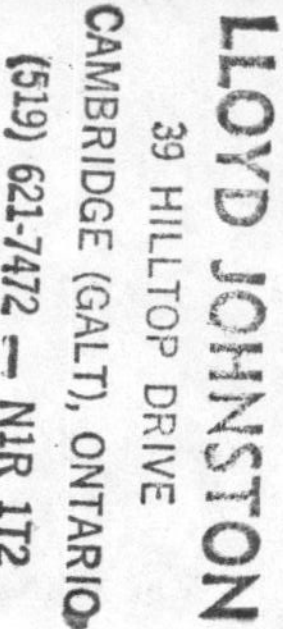




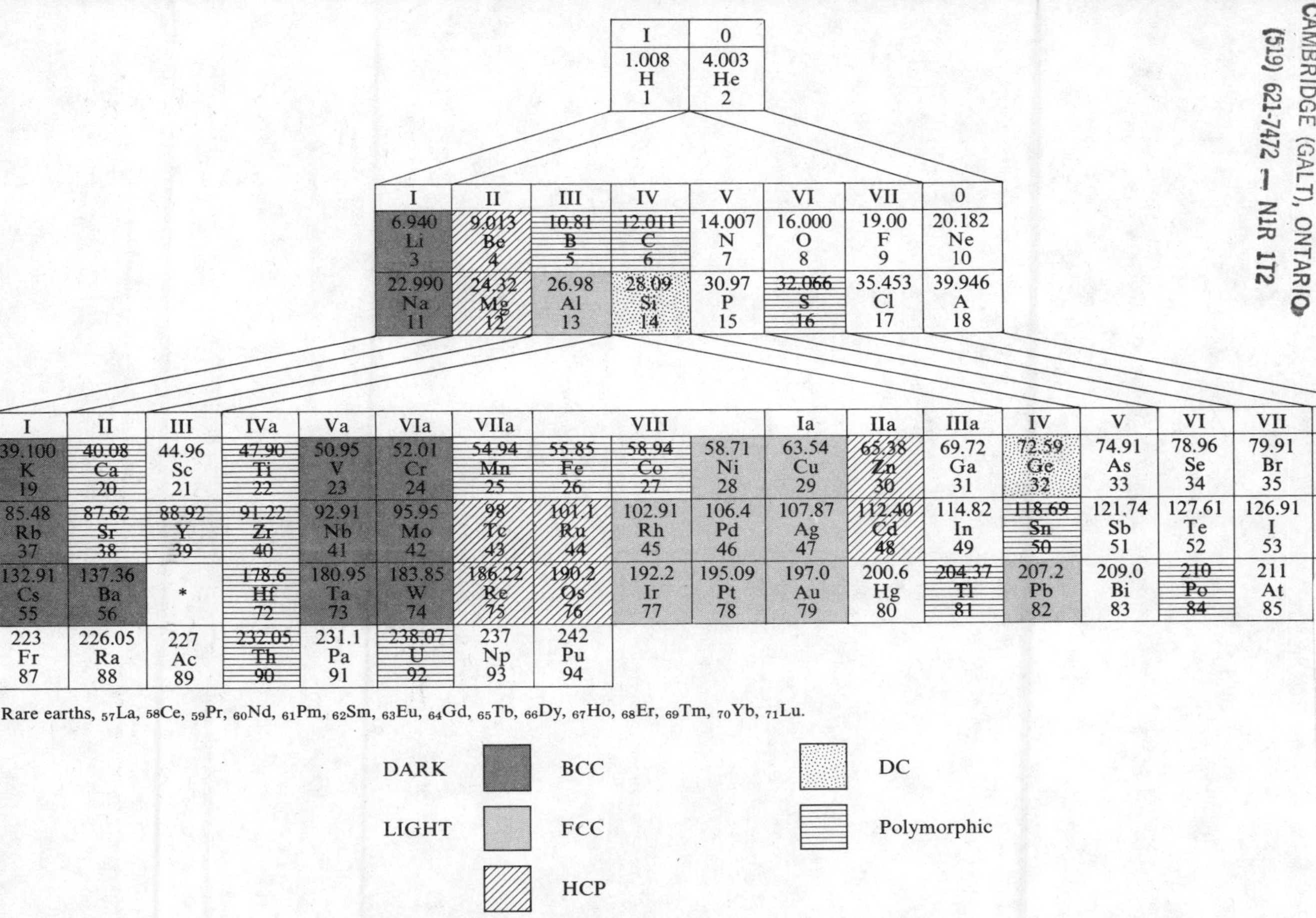

| I | 0 |
|---|---|
| 1.008 H 1 | 4.003 He 2 |

| I | II | III | IV | V | VI | VII | 0 |
|---|---|---|---|---|---|---|---|
| 6.940 Li 3 | 9.013 Be 4 | 10.81 B 5 | 12.011 C 6 | 14.007 N 7 | 16.000 O 8 | 19.00 F 9 | 20.182 Ne 10 |
| 22.990 Na 11 | 24.32 Mg 12 | 26.98 Al 13 | 28.09 Si 14 | 30.97 P 15 | 32.066 S 16 | 35.453 Cl 17 | 39.946 A 18 |

| I | II | III | IVa | Va | VIa | VIIa | | VIII | | Ia | IIa | IIIa | IV | V | VI | VII | 0 |
|---|---|---|---|---|---|---|---|---|---|---|---|---|---|---|---|---|---|
| 39.100 K 19 | 40.08 Ca 20 | 44.96 Sc 21 | 47.90 Ti 22 | 50.95 V 23 | 52.01 Cr 24 | 54.94 Mn 25 | 55.85 Fe 26 | 58.94 Co 27 | 58.71 Ni 28 | 63.54 Cu 29 | 65.38 Zn 30 | 69.72 Ga 31 | 72.59 Ge 32 | 74.91 As 33 | 78.96 Se 34 | 79.91 Br 35 | 83.80 Kr 36 |
| 85.48 Rb 37 | 87.62 Sr 38 | 88.92 Y 39 | 91.22 Zr 40 | 92.91 Nb 41 | 95.95 Mo 42 | 98 Tc 43 | 101.1 Ru 44 | 102.91 Rh 45 | 106.4 Pd 46 | 107.87 Ag 47 | 112.40 Cd 48 | 114.82 In 49 | 118.69 Sn 50 | 121.74 Sb 51 | 127.61 Te 52 | 126.91 I 53 | 131.3 Xe 54 |
| 132.91 Cs 55 | 137.36 Ba 56 | * | 178.6 Hf 72 | 180.95 Ta 73 | 183.85 W 74 | 186.22 Re 75 | 190.2 Os 76 | 192.2 Ir 77 | 195.09 Pt 78 | 197.0 Au 79 | 200.6 Hg 80 | 204.37 Tl 81 | 207.2 Pb 82 | 209.0 Bi 83 | 210 Po 84 | 211 At 85 | |
| 223 Fr 87 | 226.05 Ra 88 | 227 Ac 89 | 232.05 Th 90 | 231.1 Pa 91 | 238.07 U 92 | 237 Np 93 | 242 Pu 94 | | | | | | | | | | |

* Rare earths, $_{57}$La, $_{58}$Ce, $_{59}$Pr, $_{60}$Nd, $_{61}$Pm, $_{62}$Sm, $_{63}$Eu, $_{64}$Gd, $_{65}$Tb, $_{66}$Dy, $_{67}$Ho, $_{68}$Er, $_{69}$Tm, $_{70}$Yb, $_{71}$Lu.

# MECHANICAL BEHAVIOR

*Structure and Properties of Materials*

---

VOLUME I STRUCTURE

William G. Moffatt
George W. Pearsall
John Wulff

VOLUME II THERMODYNAMICS OF STRUCTURE

Jere H. Brophy
Robert M. Rose
John Wulff

VOLUME III MECHANICAL BEHAVIOR

Wayne Hayden
William G. Moffatt
John Wulff

VOLUME IV ELECTRONIC PROPERTIES

Leander Pease
Robert M. Rose
John Wulff

# MECHANICAL BEHAVIOR

Wayne Hayden

William G. Moffatt

John Wulff

JOHN WILEY & SONS, INC. *New York · London · Sydney*

Library of Congress Catalog Card Number: 65-14254
Printed in the United States of America

ISBN 0 471 36469 X

15 14 13 12

# *Preface*

The four brief volumes in this series were designed as a text for a two-semester introductory course in materials for engineering and science majors at the sophomore-junior level. Some curricula provide only one semester for a materials course. We have found that under such circumstances it is convenient to use Volumes I, II, and parts of III for aeronautical, chemical, civil, marine, and mechanical engineers. Similarly, parts of Volumes I, II, and IV form the basis for a single-semester course for electrical engineering and science majors.

The four volumes grew from sets of notes written for service courses during the last decade. In rewriting these for publication, we have endeavored to emphasize those principles which relate properties and behavior of different classes of materials to their structure and environment. In order to develop a coherent and logical presentation in as brief a context as possible, we have used problem sets at the end of each chapter to extend and illustrate particular aspects of the subject. Real materials encountered in engineering situations have been chosen as examples wherever possible. Problem and laboratory sections to supplement lectures have aided our students considerably in applying the principles delineated in the text to a variety of materials and environments.

Many of the tables and illustrations used in the present text have been borrowed from individual specialists and their publications. Our thanks are due both to the individuals responsible for the data and to the publishers. Their names are listed with the illustrations in the text. Further thanks are due to numerous colleagues who took part in teaching the same courses with us during the past ten years. Many parts of the text have been improved as a result of their constructive criticism.

Finally, we wish to acknowledge our indebtedness to the Ford Foundation and to Dr. Gordon S. Brown, Dean of Engineering at M.I.T., who early supported our efforts to provide lecture demonstrations, laboratory experiments, and notebook editions of the present text for the use of our students.

*January 1965*

WAYNE HAYDEN, International Nickel Company, Inc.
WILLIAM G. MOFFATT, General Electric Company
JOHN WULFF, Massachusetts Institute of Technology

# *Contents*

# MECHANICAL BEHAVIOR

CHAPTER ONE

# Mechanical Tests

The fabrication and use of materials depend in large measure on such mechanical properties as strength, hardness, and ductility. Numerical data describing these properties may be obtained from standard types of tensile, hardness, impact, creep, and fatigue tests discussed in this chapter.

## 1.1 INTRODUCTION

Since the selection of a material for a particular structural application depends on its mechanical properties, it is important to be familiar with some of the standard tests used to measure these properties and to understand the significance of the information obtained from these tests. The capacity of a material to withstand a static load can be determined by testing that material in *tension* or *compression.* Information about its resistance to permanent deformation can be gained from *hardness tests. Impact tests* are used to indicate the *toughness* of a material under shock loading conditions. When conducted over a series of temperatures, they can be used to uncover any temperature-dependent transition from ductility to brittleness. *Fatigue tests* measure a material's useful lifetime under a cyclic load. *Creep* and *stress-rupture* tests are conducted to evaluate the behavior of a material when subjected to a load at high temperatures for a long time. The results of these and more specialized tests are often of more empirical than fundamental significance. They can be, nonthe-less, useful to the designer, fabricator, and research worker.

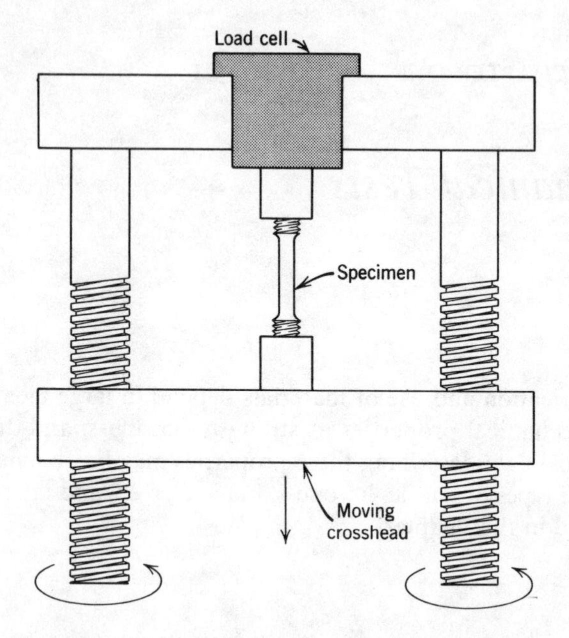


Figure 1.1 Schematic drawing of a tensile-testing apparatus.

## 1.2 THE TENSILE TEST

Of all the tests used to evaluate mechanical properties, the tensile test, in which a sample is pulled to failure in a relatively short period of time, is perhaps the most useful. In this test (see Figure 1.1) the sample is elongated in uniaxial tension at a constant rate, and the load necessary to produce a given elongation is measured as a dependent variable.

A *load-elongation* curve may be plotted from the results of a tension test. These results are usually restated in terms of stress and strain, which are independent of the geometry of the sample.

*Engineering stress,* $\sigma$, is defined as the ratio of the load on the sample, $P$, to the original cross-sectional area, $A_0$:

$$\sigma = \frac{P}{A_0} \tag{1.1}$$

*Engineering strain,* $\varepsilon$, is defined as the ratio of the change in length of the sample, $\Delta l$, to its original length $l_0$:

$$\varepsilon = \frac{l - l_0}{l_0} = \frac{\Delta l}{l_0} \tag{1.2}$$

An engineering stress-strain curve for polycrystalline copper is shown in Figure 1.2.

At the beginning of the test, the material extends elastically; this signifies that if the load is released, the sample will return to its original length. The material is said to have passed its *elastic limit* when the load is sufficient to initiate plastic, or nonrecoverable, deformation; in other words, it will no longer return to its original length if the load is released. (There is always an elastic portion of the total elongation which is recovered, but during plastic deformation a net elongation remains.) As the sample is further elongated, the engineering stress increases and the material is said to *work harden* or *strain harden.* The stress reaches a maximum at the *ultimate tensile strength.* At this point, the sample develops a neck: this is a local decrease in cross-sectional area at which further deformation is concentrated. After necking has begun, the engineering stress decreases with further strain until the sample fractures. In materials which fracture without necking, the ultimate tensile strength and the fracture strength are the same, but when necking occurs, the load at fracture is lower than the load at the ultimate tensile strength.

Figure 1.3 shows the engineering stress-strain curves for several materials. The relation between stress and strain in the elastic region is linear for metals and ceramics and is described by *Hooke's law:*

$$\sigma = E\varepsilon \tag{1.3}$$

where $E$ is a constant, *Young's modulus.* In metals and ceramics the maximum elastic strain is usually less than one-half of a percent. In rubber and other elastomers, the stress-strain relationship is nonlinear, and recoverable elastic strains of several hundred percent can be produced. The point at which deformation is no longer elastic, but plastic, is that stress at which the slope of the stress-strain curve deviates from the elastic modulus (see Figure 1.2). Because of the difficulty of determining this point precisely, various approximations are frequently used. The most common of these is the *offset yield strength,* which is usually specified as the stress at 0.2 percent plastic strain.

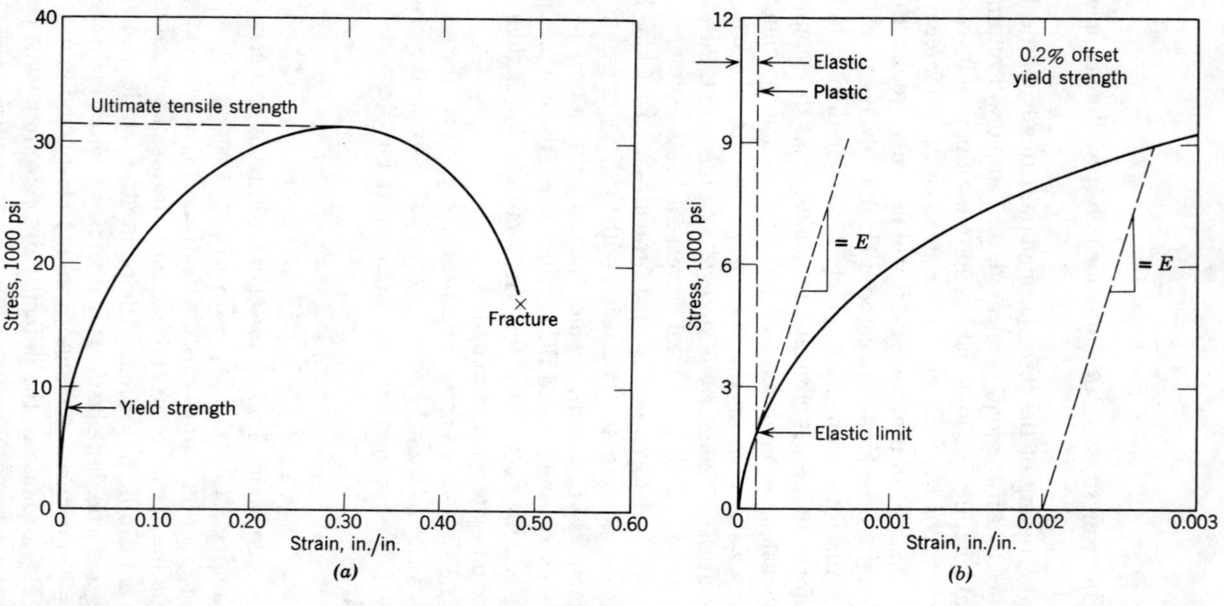


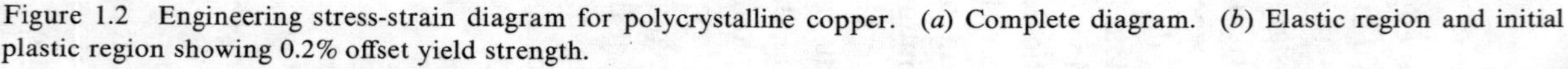

Figure 1.2 Engineering stress-strain diagram for polycrystalline copper. (*a*) Complete diagram. (*b*) Elastic region and initial plastic region showing 0.2% offset yield strength.

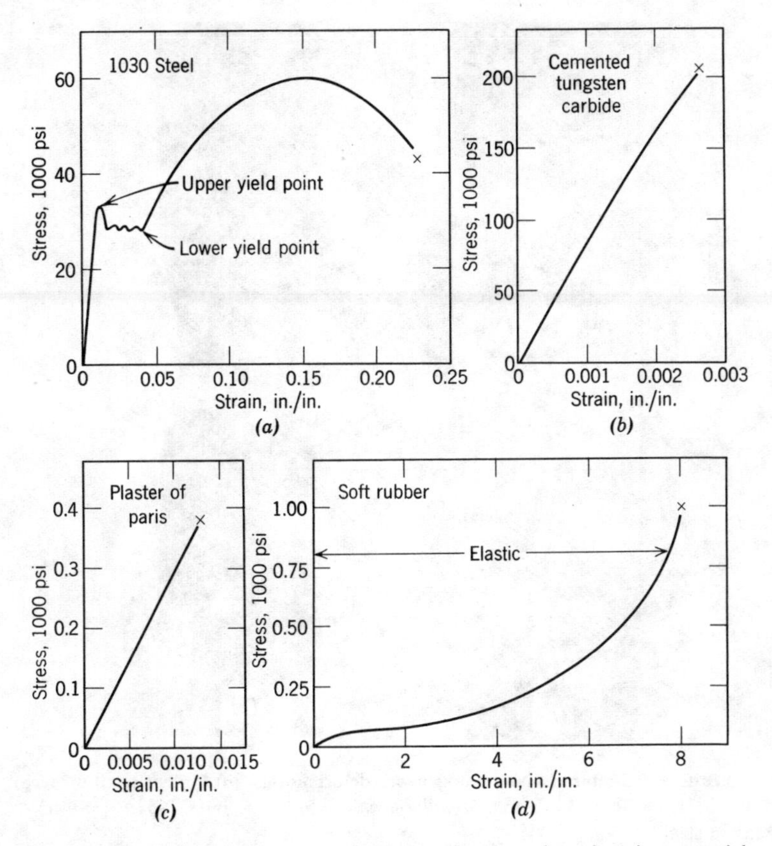


Figure 1.3 Engineering stress-strain curves for several engineering materials. (*a*) 1030 Steel. (*b*) Cemented tungsten carbide. (*c*) Plaster of Paris. (*d*) Soft rubber.

In low-carbon steels and several other alloys yielding begins at an *upper yield point,* and the stress then decreases to a *lower yield point* (see Figure 1.3*a*). This behavior stems from the nonhomogeneous deformation which begins at a point of stress concentration (often near the specimen grips) and propagates through the specimen in the form of observable bands (Lüder's bands). Figure 1.4*a* shows a sample of low-carbon steel in which Lüder's bands have partially progressed down the length of the sample.

During elastic deformation there is a slight change in the volume of a metal or ceramic sample; during plastic deformation,

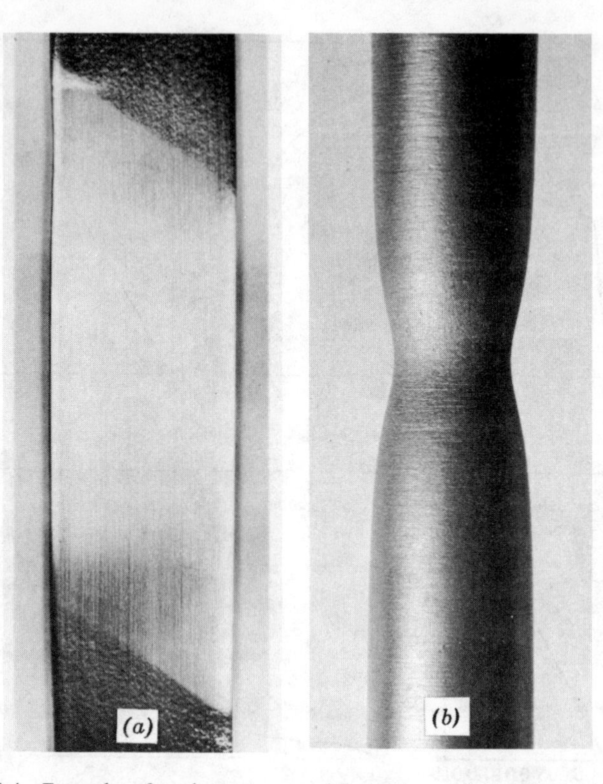


Figure 1.4 Examples of nonhomogeneous deformation. (*a*) Lüder's band in steel (1×). (From R. B. Liss, *Acta Metallurgica,* Vol. 5, No. 6, June, 1957.) (*b*) Necking in steel.

however, there is no change in the volume of the sample when measured in the unloaded state. This may be stated mathematically as

$$A_0 l_0 = A_i l_i = \text{constant} \tag{1.4}$$

When a material is elongated, its cross-sectional area must, therefore, decrease. During elastic deformation this change is negligible, but during plastic deformation the reduction in area can be appreciable. It is often preferable for this reason to redefine "stress" and "strain" when considering plastic deformation and to employ the expressions "true stress" and "true strain." *True stress is* defined as the ratio of the load on the sample to the instantaneous minimum cross-sectional area supporting that load:

$$\sigma_T = \frac{P}{A_i} \tag{1.5}$$

*True strain* is defined as the integral of the ratio of an incremental change in length to the instantaneous length of the sample:

$$\epsilon = \int_{l_0}^{l_i} \frac{dl}{l} \tag{1.6}$$

The true strain in a plastically deformed specimen of initial length $l_0$ and instantaneous length $l_i$ (measured in the unstressed state) may be calculated by integration of equation 1.6:

$$\epsilon = \ln\left(\frac{l_i}{l_0}\right) \tag{1.7}$$

If measurements are made while the sample is loaded, correction must be made for the small amount of elastic strain which always contributes to the observed elongation. This may be done with the expression:

$$(\epsilon)_{\text{loaded}} = \ln\left(\frac{l_i}{l_0}\right) - \frac{\sigma_T}{E} \tag{1.8}$$

A true stress-true strain curve for polycrystalline copper is shown in Figure 1.5. Certain fundamental differences may be observed between this curve and the plastic portion of the engineering stress-strain curve for the same material shown in Figure 1.2. True stress is not at its greatest at the initiation of necking, as is engineering stress; instead, it increases with increasing strain and reaches a maximum at fracture. Since necking is nonhomogeneous deformation, it is no longer meaningful after the beginning of necking to consider the over-all length of the sample in determining the strain. Instead, the true strain during necking

$$\epsilon = \ln\left(\frac{A_0}{A_i}\right) \tag{1.9}$$

When the increase in stress necessary to continue deforming the material is finally counterbalanced by the increase in applied stress due to the decreasing cross-sectional area, the material

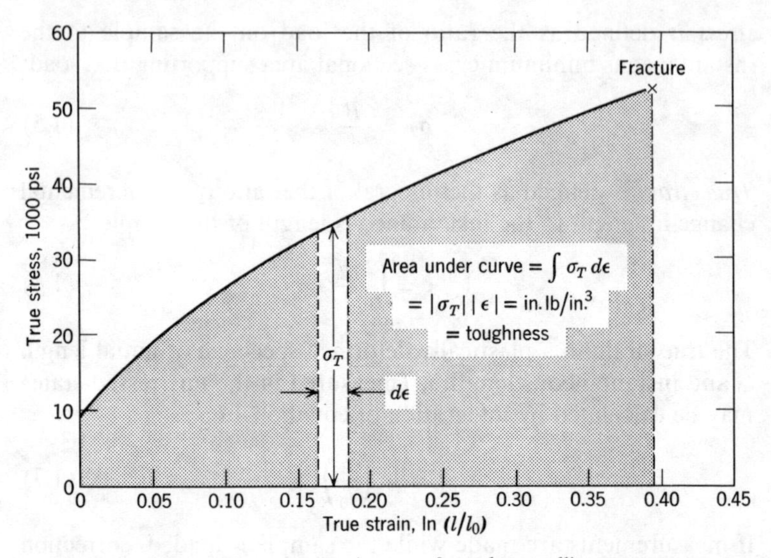


Figure 1.5 True stress-strain curve for polycrystalline copper.

necks, or starts to deform locally at decreasing loads. Mathematically, this can be expressed as:

$$\partial P = 0 = \sigma_T \, \partial A + A \, \partial \sigma_T$$
$$\frac{\partial \sigma_T}{\sigma_T} = -\frac{\partial A}{A} = \partial \epsilon \tag{1.10}$$

In many metals, work-hardening occurs in an approximately exponential manner, which may be expressed as

$$\sigma_T = K\epsilon^n \tag{1.11}$$

where $K$ is a constant and $n$ is the work-hardening exponent (always less than one). In materials whose behavior is described by Equation 1.11, necking begins when the true strain is equal to the work-hardening exponent.

The results of the tensile test are extremely useful to the designer. In most engineering structures only elastic deformation is desirable; a knowledge of the yield strength thus establishes the maximum load that can safely be employed (which is then usually reduced by a safety factor). Many techniques of fabrication, such as rolling, wire drawing, and stamping, depend on the ability of a metal to withstand appreciable plastic deformation prior to fracture. Here it is important to know the ductility and rate of work-hardening of the material being worked. The area under the stress-strain curve in the tension test shows the energy absorbed before fracture and thus indicates the *toughness* of the material.

## 1.3 THE COMPRESSION TEST

Because of the presence of submicroscopic cracks, brittle materials are often weak in tension, as tensile stress tends to propagate those cracks which are oriented perpendicular to the axis of tension. The tensile strengths they exhibit are low and usually vary from sample to sample. These same materials can nevertheless be quite strong in compression. Brittle materials are chiefly used in compression, where their strengths are much higher. Figure 1.6 shows a comparison of the compressive and tensile strengths of gray cast iron and concrete, both of which are brittle materials. A schematic diagram of a typical compression test is shown in Figure 1.7.

Because the compression test increases the cross-sectional area of the sample, necking never occurs. Extremely ductile materials are seldom tested in compression because the sample is constrained by friction at the points of contact with the platens of the apparatus. This constraint gives rise to a complicated stress distribution which can only be analyzed in an approximate fashion.

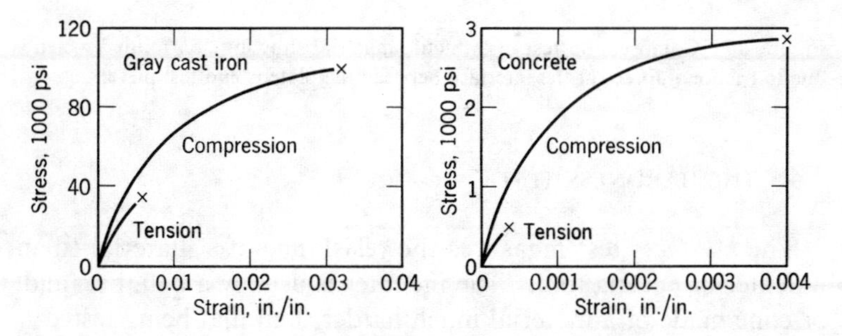


Figure 1.6 Tensile and compressive engineering stress-strain curves for gray cast iron and concrete.

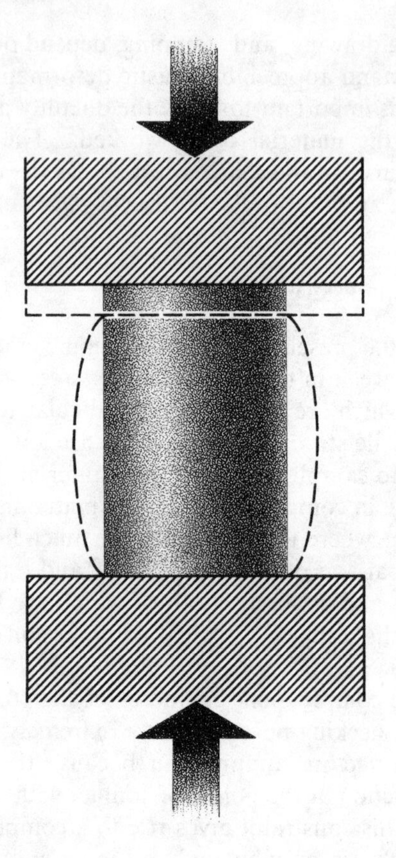

Figure 1.7 Compression test of a ductile material showing "barreling" which is due to frictional forces at the interface between the platens and test piece.

## 1.4 THE HARDNESS TEST

The *hardness* test measures the resistance of a material to an indenter or cutting tool. The indenter is usually a ball, pyramid, or cone made of a material much harder than that being tested—for example, hardened steel, sintered tungsten carbide, or diamond. In most of the standard tests, a load is applied by slowly pressing the indenter at right angles to the surface being tested for a given period of time. An empirical hardness number may be calculated from the results of such tests by knowledge of the load applied and cross-sectional area or depth of the impression (see Table 1.1). These tests are never taken near the edge of a sample or any closer than three times the diameter of an impression to an existing impression. The thickness of the test specimen should be at least ten and one-half times the depth of the impression. Microscopic hardness indentations are made with very small loads. They are used to study local hardness variations in single-phase and multiphase materials.

Most hardness tests produce plastic deformation in the material, and all variables which affect plastic deformation affect hardness. For materials which work-harden in a similar fashion, there is a good correlation between hardness and the ultimate tensile strength. The hardness test may be conducted easily, and the information obtained from it is readily evaluated. For these reasons and because it is nondestructive, it is frequently employed for quality control in production.

## 1.5 THE IMPACT TEST

The *impact test* measures the energy necessary to fracture a standard notched bar by an *impulse* load and as such is an indication of the *notch toughness* of a material under shock loading. Figure 1.8 shows the apparatus and sample geometry for one of the standard impact tests. The sample is placed across parallel jaws of the testing mechanism. A heavy pendulum, released from a known height, strikes and breaks the sample before it continues its upward swing. From knowledge of the mass of the pendulum and the difference between the initial and final heights, the energy absorbed in fracture can be calculated. The presence of the notch in the bar and the almost instantaneous nature of the loading increase the severity of the test, as will be shown in Chapter 7. The stress concentration at the base of the notch produces fracture with little plastic flow.

The impact test indicates the notch sensitivity of a material resulting from the presence of internal stress raisers, such as grain

*Table 1.1 Hardness Tests*

| Test | Indenter | Shape of Indentation: Side View | Shape of Indentation: Top View | Load | Formula for Hardness Number |
|---|---|---|---|---|---|
| Brinell | 10 mm sphere of steel or tungsten carbide | $D$, $d$ | $d$ | $P$ | $\text{BHN} = \dfrac{2P}{\pi D[D - \sqrt{D^2 - d^2}]}$ |
| Vickers | Diamond pyramid | 136° | $d_1$, $d_1$ | $P$ | $\text{VHN} = 1.72P/d_1^2$ |
| Knoop microhardness | Diamond pyramid | $t$; $l/b = 7.11$; $b/t = 4.00$ | $b$, $l$ | $P$ | $\text{KHN} = 14.2P/l^2$ |
| Rockwell | | | | | |
| A | Diamond cone | 120°, $t$ | | 60 kg | $R_A =$ 100–500$t$ |
| C | Diamond cone | | | 150 kg | $R_C =$ 100–500$t$ |
| D | Diamond cone | | | 100 kg | $R_D =$ 100–500$t$ |
| B | $\frac{1}{16}$ in. diameter steel sphere | $t$ | | 100 kg | $R_B =$ 130–500$t$ |
| F | $\frac{1}{16}$ in. diameter steel sphere | | | 60 kg | $R_F =$ 130–500$t$ |
| G | $\frac{1}{16}$ in. diameter steel sphere | | | 150 kg | $R_G =$ 130–500$t$ |
| E | $\frac{1}{8}$ in. diameter steel sphere | | | 100 kg | $R_E =$ 130–500$t$ |

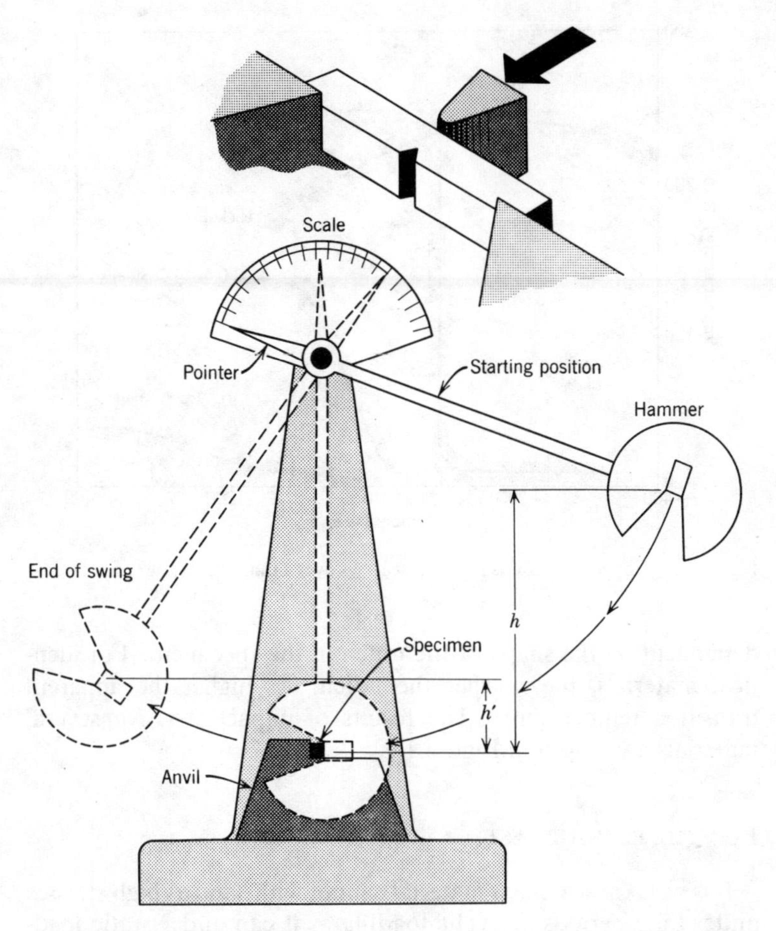


Figure 1.8 Schematic drawing of a standard impact-testing apparatus.

boundary inclusions, internal cracks, and second phases. It is also useful as a production tool in comparing manufactured materials with others which have proved satisfactory in service. Steels, like most other BCC metals and alloys, absorb more energy when they fracture in a ductile fashion rather than in a brittle fashion. On this account the impact test is often used to assess the temperature of the *transition* from the ductile to brittle state which occurs as the temperature is lowered. *The transition temperature is* also

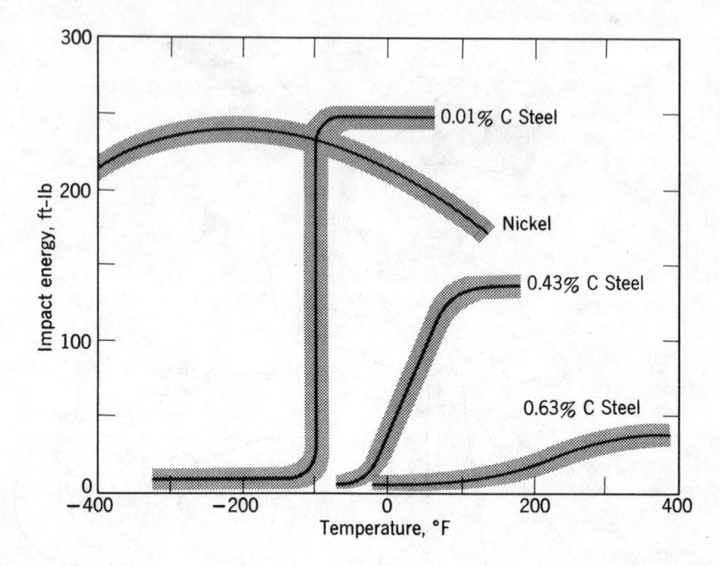


Figure 1.9 Impact test results for several alloys over a range of testing temperatures.

dependent on the shape of the notch in the specimen. For identical materials, the sharper the notch, the higher the apparent transition temperature. The results of impact tests for several materials are shown in Figure 1.9.

## 1.6 THE FATIGUE TEST

It is well known that a material cannot withstand as high stresses under long periods of cyclic loading as it can under static loading. The yield strength and the ultimate tensile strength are useful measures of the load-carrying ability of materials operating under static loads only. The *fatigue* test determines the stresses which a sample of a material of standard dimensions can safely endure for a given number of cycles. In one common fatigue test, the apparatus for which is shown in Figure 1.10, the sample is loaded in pure bending. The sample is then rotated, and with each rotation all points on the circumference of the sample pass from a state of compression to one of tension. Each revolution thus constitutes a complete cycle of stress reversal which in the

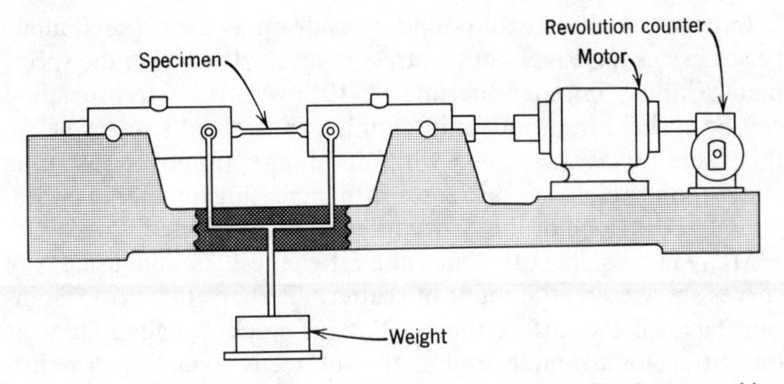


Figure 1.10 Schematic diagram of a R. R. Moore reversed-bending fatigue machine.

usual test is repeated several thousand times per minute. Specimens are tested to failure using different loads, and the number of cycles before failure is noted for each load. The data are then plotted as the stress versus the logarithm of the number of cycles to failure. Examples of this type of curve (*S-N* plot) for steel and aluminum are shown in Figure 1.11.

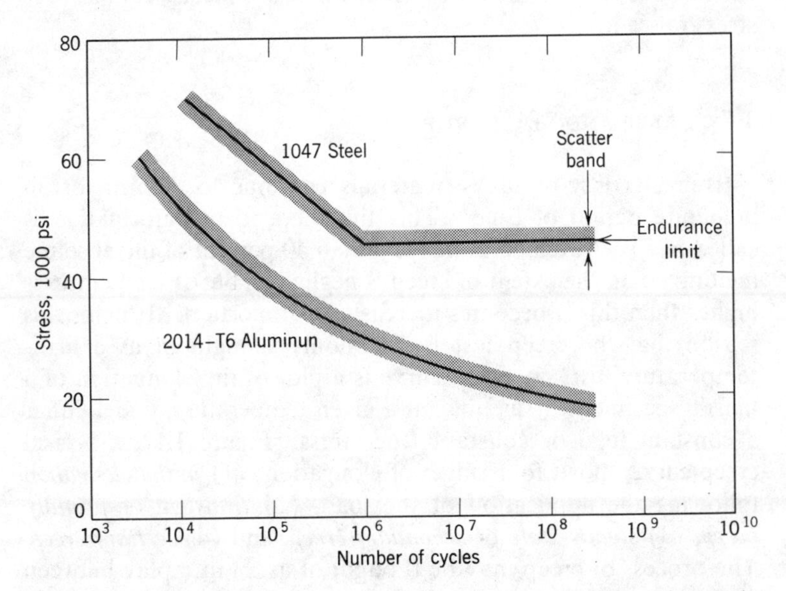


Figure 1.11 *S-N* curves for fatigue failure for aluminum and low-carbon steel.

In ferrous alloys the allowable stress drops as the required number of cycles increases until a stress is reached at which the specimen exhibits "infinite" lifetime ($>10^8$ cycles). A ferrous alloy can be cycled for an indefinite number of times at stresses below this value (*endurance limit*) without failing. In most nonferrous alloys the stress always decreases with increasing number of cycles, although the slope of the *S-N* curve usually decreases.

Many factors affect the data in a fatigue test, and the usual *S-N* curves show a great amount of scatter. Perhaps the most significant factor is the surface finish. Fatigue cracks usually initiate at the surface of a sample, and as the surface is given a finer polish the level of the *S-N* curve (and the endurance limit of ferrous alloys) increases. The fatigue strength is increased if the surface of the sample is hardened by chemical or mechanical techniques which produce surface compressive stresses. The chemical nature of the environment also has an effect; fatigue strengths are generally lower in a corrosive environment than in a noncorrosive one. If a steady stress is applied while the sample is being alternated in tension and compression, the fatigue life is changed. A steady tensile stress lowers the fatigue life, whereas a steady compressive stress raises it.

## 1.7 CREEP AND STRESS RUPTURE

Even at constant stress, materials continue to deform for an indefinite period of time. This time-dependent deformation is called *creep*. At temperatures less than 40 percent of the absolute melting point, the extent of creep is negligible, but at temperatures higher than this it becomes increasingly important. It is for this reason that the creep test is commonly thought of as a high-temperature test. A creep curve is a plot of the elongation of a tensile specimen versus time, at a given temperature, under either a constant load or constant true stress. Figure 1.12, a typical creep curve, shows four stages of elongation: (1) *initial elongation* following the application of the load, (2) *transient* or *primary creep*, (3) *steady-state* or *secondary creep*, and (4) *tertiary creep*. The process of creep may be thought of as an interplay between two mechanisms which we shall study in Chapter 6. At present

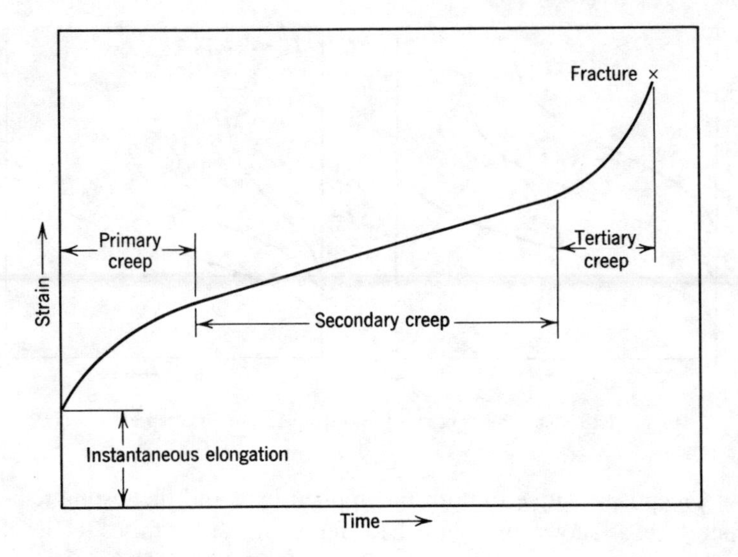


Figure 1.12 Typical creep curve showing the four stages of elongation for a long-time, high-temperature creep test.

we shall state that they are work-hardening and softening due to recovery processes, such as dislocation climb, thermally activated cross-slip, and vacancy diffusion. In transient creep the rate of work-hardening is faster than the rate of recovery, and thus the creep rate continually decreases. During steady-state creep the rates of work-hardening and recovery are approximately equal, and the creep rate is nearly constant. There are several factors which can lead to tertiary creep. Cracks may begin to form at grain boundaries in the sample and thus diminish its effective load-carrying, cross-sectional area. Necking may begin at some point in the sample and decrease its actual load-carrying, cross-sectional area; a softening process may begin to proceed at a higher rate than work-hardening. All these factors will produce an ever-increasing rate of elongation until the sample fractures. In materials which neck appreciably, the occurrence of tertiary creep can be eliminated by the application of a constant true stress instead of a constant load. However, even if a constant true stress is applied, tertiary creep is observed if grain boundary cracking or a rapid recovery process occurs.

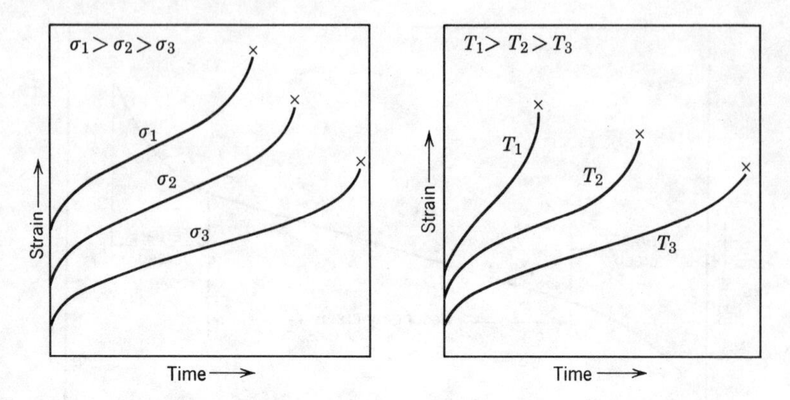


Figure 1.13 The effects of stress and temperature on creep behavior.

Creep is sensitive to both the applied load and the testing temperature, as shown in Figure 1.13: increasing stress raises the level of the creep curve, and increasing temperature, which accelerates recovery processes, increases the creep rate.

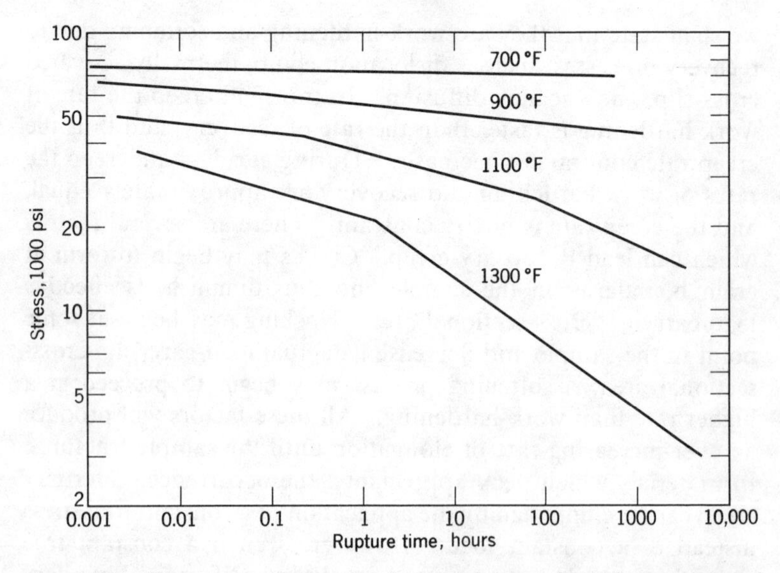


Figure 1.14 Stress versus rupture time for annealed monel tested over a range of temperatures. (After N. J. Grant and A. G. Bucklin, *Trans, ASM,* 1953, p. 156.)

The *stress-rupture* test is an extension of the creep test. In this test a sample is held under an applied load at a definite temperature until it fractures. The elongation of the sample, the time until fracture, the applied load, and the testing temperature are all recorded. From these data several curves can be plotted, all of which are helpful to the designer. Curves relating the applied stress to the rupture time at various temperatures are shown in Figure 1.14.

## DEFINITIONS

*Engineering Stress.* The load on a sample divided by the original cross-sectional area.

*Engineering Strain.* The change in length of a sample divided by the original length.

*True Stress.* The load on a sample divided by the instantaneous cross-sectional area.

*True Strain.* The natural logarithm of the ratio of the instantaneous length of a plastically deformed sample to its original length; or, the natural logarithm of the ratio of the original cross-sectional area of such a sample to its instantaneous cross-sectional area.

*Elastic Deformation.* The change of shape of a stressed body which is recovered when the stress is released.

*Plastic Deformation.* The change of shape of a stressed body which is not recovered when the stress is released.

*Elastic Limit.* That stress at which a material deviates from linear elastic behavior.

*Yield Strength.* That stress needed to produce a specified amount of plastic deformation (usually, an 0.2 percent change in length).

*Ductility.* The amount of plastic strain that a material can withstand before fracture.

*Necking.* The concentration of plastic deformation in a localized region of a sample under tension.

*Ultimate Tensile Strength.* The maximum engineering stress which a material can withstand.

*Work-Hardening, Strain-Hardening.* The increase in hardness and flow stress occurring with increasing plastic deformation.

*Fatigue Strength.* The maximum cyclic stress a material can withstand for a given number of cycles before failure occurs.

*Endurance Limit.* The stress below which a material can be stressed cyclically for an infinite number of times without failure.

*Creep.* The plastic deformation of a material which occurs as a function of time when that material is subjected to a constant stress or load.

## BIBLIOGRAPHY

SUPPLEMENTARY READING:

*American Society for Metals Handbook,* Vol. I, 8th edition, Novelty, Ohio, 1961, pp. 661, 662.

*American Society for Testing Materials,* Philadelphia, 1958.
ASTM Standard E 10-58T 3, 30–38
ASTM Standard E 18-57T 3, 39–51
ASTM Standard D 1415-56T 9, 1299–1382

Cowdrey, I. H. and R. G. Adams, *Materials Testing,* 2nd edition, John Wiley and Sons, New York, 1944, pp. 5–81.

Dieter, G. E., *Mechanical Metallurgy,* Vol. 3, McGraw-Hill Book Co., New York, 1961, Chapters 9–16.

Marin, J., *Mechanical Behavior of Engineering Materials,* Prentice-Hall, Englewood Cliffs, N.J., 1952, Appendix B, pp. 475–483; Chapter 10, pp. 446–463.

McClintock, F. A. and A. S. Argon, *Introduction to the Mechanical Behavior of Materials,* M.I.T., Cambridge, Mass., 1962, Chapters 1, 2, 8, 9, 23.

McLean, D., *Mechanical Properties of Metals,* John Wiley and Sons, New York, 1962, Chapter 4, pp. 97–109; Chapter 5, pp. 112–159.

ADVANCED READING

Alfrey, T. Jr., *Mechanical Behavior of Polymers,* Interscience Publishers, New York, 1948.

Ferry, J. D., *Viscoelastic Properties of Polymers,* John Wiley and Sons, New York, 1960.

O'Neil, H., "Significance of Tensile and Other Mechanical Tests," *Proc. Institute of Mechanical Engineers,* **151** (1944) 116–150.

Tabor, D., *The Hardness of Metals,* Oxford University Press, London, 1951.

Westbrook, J. H., "Temperature Dependence of the Hardness of Pure Metals", Trans. A.S.M. **45** (1953), 221–248.

## PROBLEMS

1.1 A steel bar and an aluminum bar are each under a 1000-lb load. If the cross-sectional area of the steel bar is one square inch, what must the cross-sectional area of the aluminum bar be for the elastic strains in both bars to be the same? (Steel: Young's modulus $= 30 \times 10^6$ psi; aluminum: Young's modulus $= 10 \times 10^6$ psi.)

1.2 (a) What is meant by a 0.2 percent yield strength?

(b) What is meant by 0.002 offset in the tensile test?

1.3 For a material whose behavior under plastic tensile deformation is described by Equation 1.11, show that necking begins when the true strain, $\epsilon$, is equal to the strain-hardening exponent, $n$.

1.4 Show that for small plastic strains, the engineering strain and the true strain are nearly equal. (*Hint:* this problem may be done by a series expansion of a natural logarithm.)

1.5 Plastic deformation of annealed 24-S aluminum follows the relationship expressed in Equation 1.11. It has been found experimentally that $K = 49{,}000$ psi and that $n = 0.21$. From these data calculate the 0.2 percent offset yield strength of this alloy.

1.6 Show that the true stress during plastic deformation can be related to the engineering stress by the equation:

$$\sigma_T = \sigma(1 + \varepsilon)$$

1.7 Explain why the level of the engineering stress-strain curve is lower than that of the true stress-strain curve in the tension test, while the opposite is true in the compression test.

1.8 When a ductile material is tested in compression it develops a barrel shape, that is, the diameter in the middle of the sample is larger than the diameter at the ends of the sample. Discuss why this occurs.

1.9 Cemented carbides, cast iron, ceramics, and other brittle materials usually fracture in tension with little or no observable plastic deformation. These materials are difficult to machine into tensile specimens, and more difficult to grip and load without inducing stress concentrations. Instead they are commonly tested as a simply supported beam with a span $l$ between supports, loaded at mid-span. The deflection at mid-span is then measured as a function of applied load.

(a) For a bar of square cross section $a^2$, with a distance $l$ between supports, show that if little yielding occurs, the breaking strength, $S$, is related to the breaking load, $F$, by the equation

$$S = \left(\frac{3l}{2a^3}\right)F$$

(b) Derive a similar equation for the breaking strength of a bar of radius $r$.

1.10 Why is the hardness of materials of interest to each of the following: (a) the machinist, (b) the design engineer, (c) the mineralogist, (d) the testing engineer.

1.11 What effect will each of the following conditions have on the observed hardness value of a material?

(a) An indentation made too close to an existing indentation.

(b) An indentation made too close to the edge of the sample.

(c) An indentation on a sample whose thickness is less than ten and one-half times the depth of the indentation.

1.12 The hardness of a cylindrical sample of a material can be tested by placing the sample in a *V-block* holder. Explain why the hardness number decreases as the radius of samples of otherwise identical material decreases.

1.13 What is the ratio of the mean compressive stress beneath a hardness indenter during hardness testing to the stress necessary to induce plastic flow in a conventional compression or tensile test? Why the difference? How do the two tests differ in amount of strain?

1.14 (a) Define cutting hardness.

(b) List specific energy associated with cutting for lead, magnesium, aluminum, copper, cast iron, steel (structural), and steel (tool).

1.15 Discuss why the *ductile-to-brittle transition temperature* measured by impact testing decreases as the radius of curvature of the notch in the sample increases.

1.16 For a given steel in various stages of heat treatment, the endurance limit of a specimen with a sharp notch as compared to that of a smoothly polished specimen decreases linearly from 85 percent at an ultimate tensile strength of 40,000 psi to 35 percent at 220,000 psi. Assuming the endurance limit of a smoothly polished specimen is one-half the ultimate tensile strength, to what strength level should the steel be heat-treated in order that a notched part would show the greatest endurance limit?

1.17 During a creep test, continuing measurements of minimum specimen diameter are made and the load is adjusted to maintain a constant ratio of $4P/\pi D_0^2$. Nonetheless, a pronounced region of tertiary creep is observed. Explain this observation.

1.18 If a certain steel fails in a brittle manner in an impact test at a temperature of 0° C, does this mean that any structure made of this same steel will fail in a brittle manner at this temperature? Why?

CHAPTER TWO

# *Elastic Properties*

Elastic deformation of solids is limited. The strain induced in them by a given stress completely disappears when the stress is removed. The relationship between stress and strain, which is linear in some materials but highly nonlinear in others, may be correlated qualitatively with the structure and type of atomic bonding present. This stress-strain relationship also depends on temperature and, in single crystals, or specifically worked materials, on crystallographic direction.

## 2.1 INTRODUCTION

All materials change in shape, volume, or both under the influence of an applied stress or a temperature change. The deformation is called *elastic* if the stress- or temperature-induced change in shape or volume is completely recovered when the material is allowed to return to its original temperature or state of stress. In crystalline substances the relationship between stress and strain in the elastic region is typically linear, whereas noncrystalline long-chain molecular materials generally exhibit nonlinear elastic behavior. Two graphs of stress ($\sigma$) versus strain ($\varepsilon$) are shown in Figure 2.1.

The simple mathematical theory of *linear elasticity* considerably antedates any detailed knowledge of the atomic basis of the behavior observed and merely deals with the proportionality between stress and strain on a macroscopic scale, utilizing elastic constants which can be determined by tests of the sort described in Chapter 1. For example, the equation for the linear portion of the tensile stress-strain curve (Figure 2.1*a*) is $\sigma = E\varepsilon$, where $E$ is the

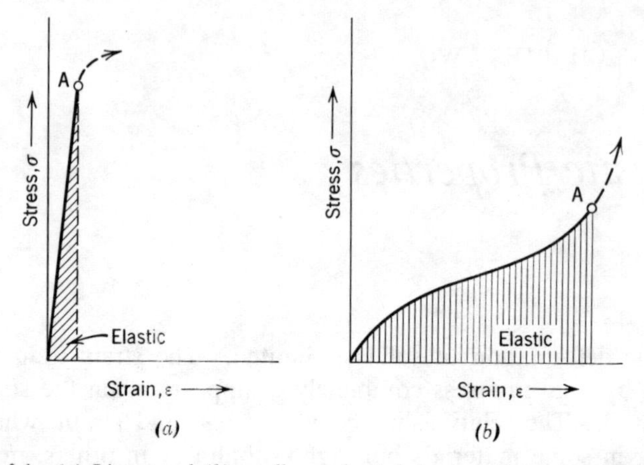


Figure 2.1 (*a*) Linear and (*b*) nonlinear elastic behavior. Point A on each curve represents the end of the elastic region.

proportionality constant, Young's modulus. The value of Young's modulus may be determined by other means: for example, if $v$ is the velocity of sound in a material of density $\rho$, and Young's modulus $E$, then

$$v = \sqrt{E/\rho} \tag{2.1}$$

Several different elastic proportionality constants are in common use; they differ only in the types of stress and strain which they relate:

Young's modulus $$E = \frac{\sigma}{\varepsilon} \tag{2.2}$$

Shear modulus $$G = \frac{\tau}{\gamma} \tag{2.3}$$

Bulk modulus $$K = \frac{\sigma_{\text{Hyd}}}{\Delta V/V_0} \tag{2.4}$$

In the above equations, $\sigma$ is uniaxial tensile or compressive stress, $\tau$ is shear stress, $\sigma_{\text{Hyd}}$ is *hydrostatic* tensile or compressive stress, $\varepsilon$ is normal strain, $\gamma$ is shear strain and $\Delta V/V_0$ is fractional volume expansion or contraction. Figures 2.2, 2.3, and 2.4 illustrate the geometric relationship between stress and strain of the types described in the three equations above.

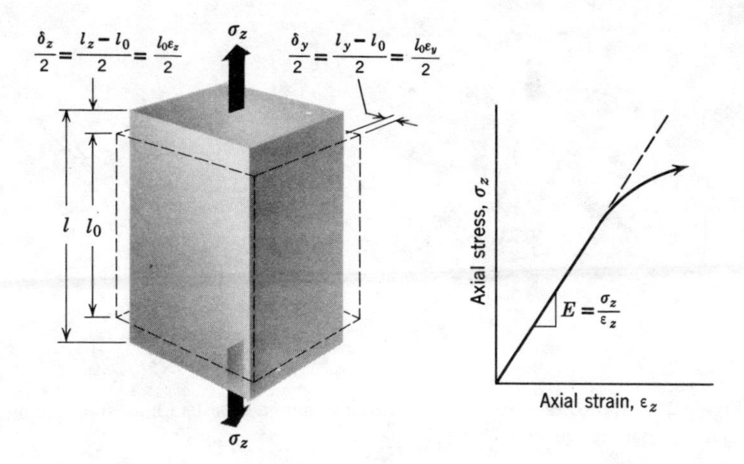


Figure 2.2 Uniaxial tensile (or compressive) stress. Poisson's ratio, $\nu$, is the ratio of transverse to axial strain. Dashed lines represent initial stress-free shape: a cube of edge length $l_0$.

*Poisson's ratio*, $\nu$, another elastic constant, is the ratio of transverse to axial strain (see Figure 2.2):

$$\nu = \frac{-\varepsilon_y}{\varepsilon_z} \tag{2.5}$$

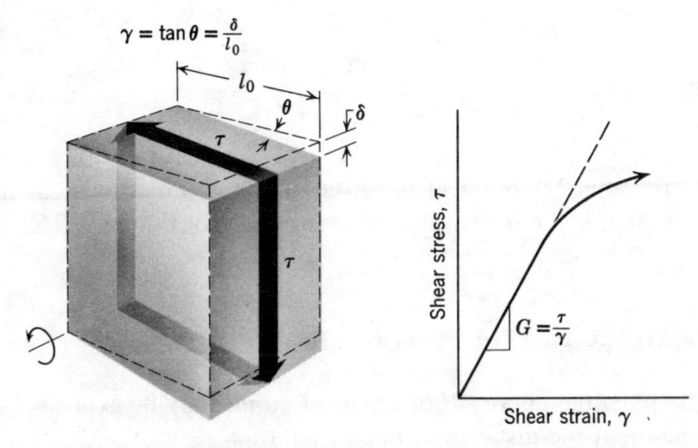


Figure 2.3 Geometry of shear stress–shear strain relationship. Dashed lines represent initial stress-free shape: a cube of edge length $l_0$. (A rigid body rotation, as indicated, would also occur.) Shear strain $\gamma = \delta/l_0 = \tan\theta$.

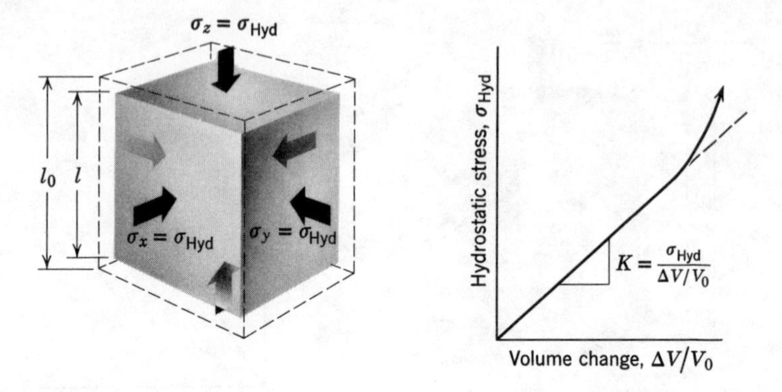


Figure 2.4 Hydrostatic stress versus volume change. Dashed lines surrounding cube represent initial stress-free size.

In isotropic elasticity, if any two of the four elastic constants, $E$, $G$, $K$, and $\nu$, are known for a material which is homogeneous (in which the properties do not vary from point to point) and isotropic (in which the properties at a point are identical in all directions) the other two may be derived:

$$K = \frac{E}{3(1 - 2\nu)} \tag{2.6}$$

$$G = \frac{E}{2(1 + \nu)} \tag{2.7}$$

$$\nu = \frac{E}{2G} - 1 \tag{2.8}$$

See Problems 2.9 and 2.11 for a derivation of Equations 2.6 and 2.7. Equation 2.8 is merely a rearrangement of Equation 2.7.

## 2.2 ATOMIC BASIS OF ELASTIC BEHAVIOR

The potential energy $V$ of a pair of atoms may be expressed as a function of the distance of their separation $r$:

$$V = \frac{-A}{r^n} + \frac{B}{r^m} \tag{2.9}$$

where $A$ and $B$ are, respectively, the proportionality constants for attraction and repulsion and $n$ and $m$ are exponents giving the appropriate variation of $V$ with $r$. Expressions for the forces of attraction and repulsion existing between the two atoms may be derived from the expression for potential energy, in the form:

$$F = \frac{-\partial V}{\partial r} = \frac{-nA}{r^{n+1}} + \frac{mB}{r^{m+1}} \tag{2.10}$$

Letting $nA = a$, $mB = b$, $n + 1 = N$ and $m + 1 = M$:

$$F = \frac{-a}{r^N} + \frac{b}{r^M} \tag{2.11}$$

Plots of the curves (*Condon-Morse* curves) of Equations 2.9 and 2.11, which, clearly, have the same form, are shown in Figure 2.5. The value of $r$ corresponding to the minimum of potential energy is the equilibrium spacing, $d_0$, of the two atoms. The net force is zero at $d_0$, and a displacement in either direction will call restoring forces into play. Although these curves describe the behavior of an isolated atom pair, the same kind of behavior is exhibited as a free atom approaches an existing crystal lattice: a net attractive force at first exists (potential energy decreases)

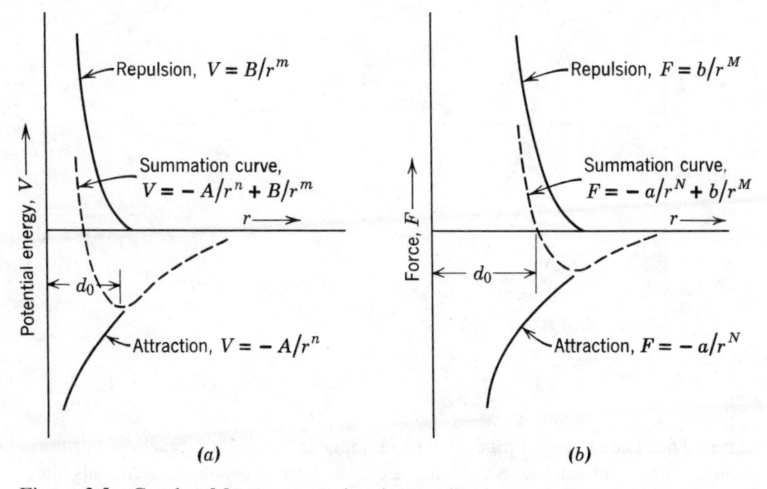


Figure 2.5 Condon-Morse curves showing qualitative variation of (*a*) energy and (*b*) force with distance of separation, $r$, of atoms.

which then reduces to zero (potential energy reaches a minimum) at a distance $d_0$, where the forces of attraction and repulsion are in balance. If the interatomic spacing were further reduced, a net repulsive force would act to restore the atoms to their equilibrium spacing. Atoms in a crystal structure tend, therefore, to be arrayed in a definite pattern with respect to their neighbors.

Macroscopic elastic strain results from a change in interatomic spacing. The macroscopic strain, $(l - l_0)/l_0$, in a given direction is equal to the average fractional change in interatomic spacing, $(d - d_0)/d_0$, in that direction. It may easily be shown, then, that Young's modulus $E$ is proportional to the slope, at $d_0$, of the Condon-Morse force curve or, alternatively, to the curvature of the Condon-Morse potential curve at $d_0$. The normal range of elastic strain in crystalline materials rarely exceeds $\pm\frac{1}{2}$ percent. Since, as may be seen in Figure 2.6, the tangent $\partial F/\partial r$ very nearly

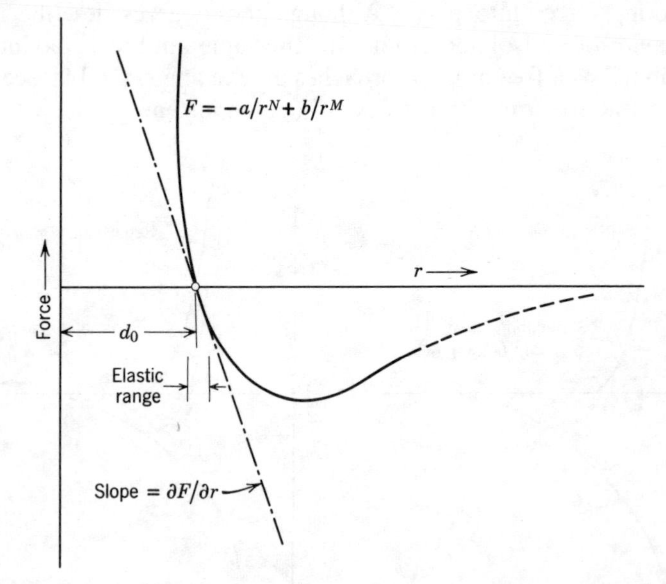


Figure 2.6 The summation curve and its tangent are, for all practical purposes, coincident over the range of elastic strains encountered in crystalline materials; thus, with virtually no error, stress may be considered proportional to strain in the elastic range.

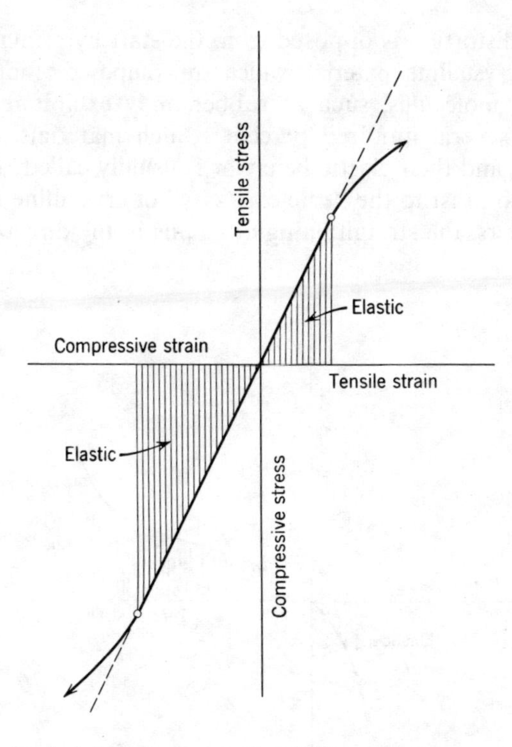


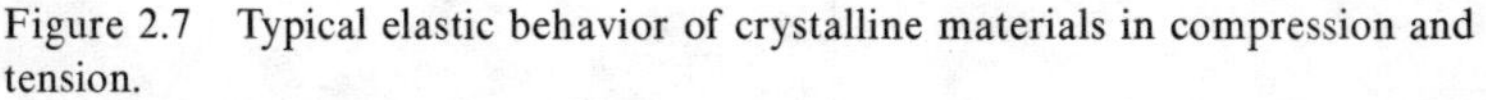

Figure 2.7 Typical elastic behavior of crystalline materials in compression and tension.

coincides with the force curve in this area of strain, it is clear that for all practical purposes stress is, as the theory of elasticity states, a linear function of strain.

Although the maximum elastic strain in crystalline materials is usually very small, the stress necessary to produce this strain is usually great. This stress-strain ratio is high because the applied stress works in opposition to the restoring forces of primary bonds (ionic, covalent, metallic). The elastic behavior of such materials under compression is the same as their behavior under tension, and the compressive stress-strain curve is merely an extension of the tensile stress-strain curve, as is shown in Figure 2.7.

Certain noncrystalline materials, such as glass or cross-linked polymers, may also exhibit linear elasticity, for their structure is such that distortion is opposed from the start by primary bonds. Other noncrystalline materials which are composed of intertangled long-chain molecules, such as rubber, may exhibit recoverable strains of several hundred percent. Such materials are called *elastomers,* and their elastic behavior is usually called "high elasticity" in contrast to the "true elasticity" of crystalline materials. In elastomers, the straightening of chains in the direction of the

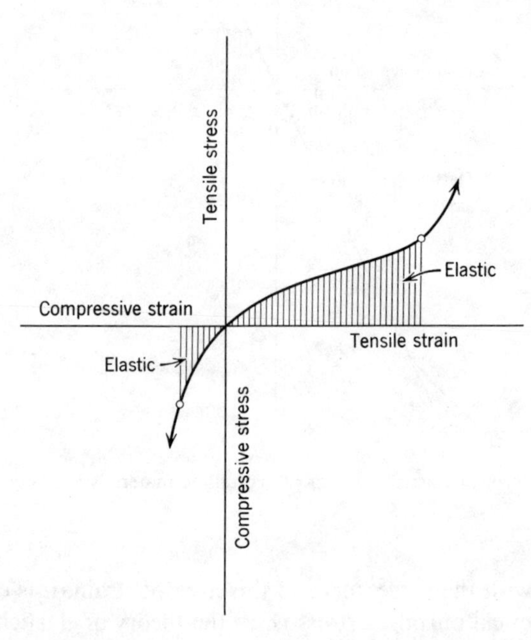


Figure 2.8 Typical elastic behavior of elastomers in compression and tension.

applied stress can produce appreciable macroscopic elastic strain at low stresses. Once the chains have been aligned, however, further elastic elongation requires the stretching of the chains in opposition to the primary bonding forces within them, and to the secondary bonding forces between them. Elastomers therefore show the nonlinear elastic tensile behavior illustrated in Figure 2.1*b* and again in Figure 2.8. Compressive stress applied to elastomers (see Figure 2.8) initially causes a more efficient filling of space in the material. As the available space decreases, the resistance to further compression increases, until finally the primary bonding forces within the chains begin to oppose the applied stress. The stress-strain curve in compression thus increases in slope as deformation increases.

Certain *cellular* substances, such as wood, may be fairly stiff in compression until the stress is sufficient to cause elastic buckling

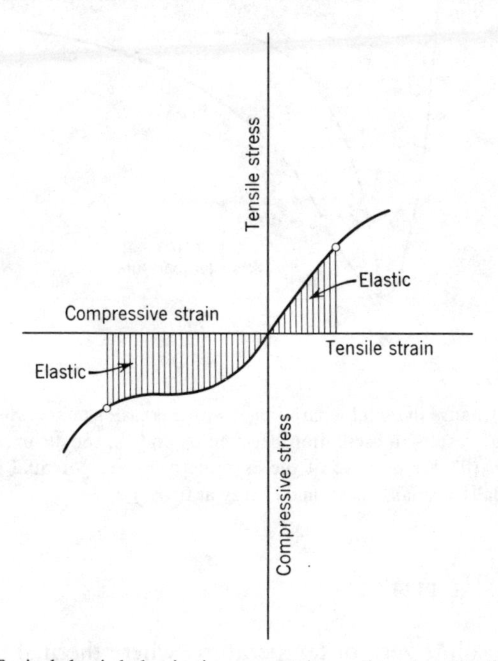


Figure 2.9 Typical elastic behavior in compression and tension of cellular materials that exhibit elastic buckling of cell walls under compression.

of the cell walls, at which point considerable strain may accumulate without much increase in stress. The stiffness may then increase again as the cells become compacted. Very considerable nonlinear strains may be recoverable in such substances; of course, if the stress becomes high enough, the cells will crush, and the strain may not then be recovered. In tension, clearly, the cell walls do not buckle elastically in the same way; a typical stress-strain curve is shown in Figure 2.9.

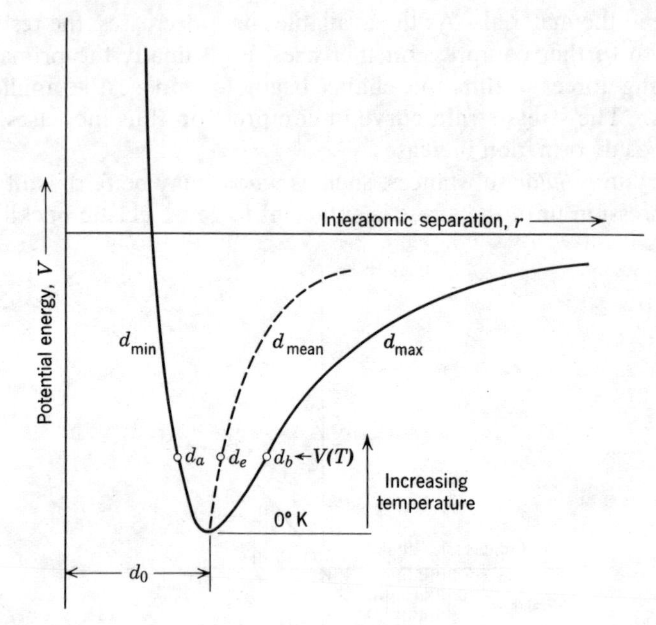


Figure 2.10 Change in radial separation, $r$, with increasing temperature. At $T_1$ the thermal energy results in oscillation between $d_a$ and $d_b$; the mean position, $d_e$, is greater than $d_0$(0° K) because of the asymmetry of the potential energy curve. (There is actually a small vibrational energy at 0° K.)

## 2.3 THERMAL EFFECTS

At the absolute zero of temperature, where thermal motion has ceased, the interatomic spacing has the value $d_0$ (see Figure 2.10). As thermal energy is added to the two atoms, however, they begin to oscillate about their equilibrium positions, $d_e$. Thus, at the temperature $T_1$ in Figure 2.10, the minimum and maximum interatomic spacings are at points $d_a$ and $d_b$, respectively, and the mean interatomic spacing is at point $d_e$. Clearly, because of the assymetry of the potential energy curve, the mean spacing will increase with increasing temperature (except during some allotropic transformations). The linear (unidirectional) thermal expansion resulting from a temperature increase, $dT$, may be expressed by

$$\frac{dl}{l} = \alpha\, dT \tag{2.12}$$

where $l$ is the length at a given temperature, $dl$ is the change in length resulting from temperature change $dT$, and $\alpha$ is the *linear thermal expansion coefficient* (which is a function of temperature). Thermal expansion is usually isotropic in cubic crystals, and a single linear thermal expansion coefficient is adequate to describe the volume expansion (three times the linear expansion) on heating. In materials of other crystal structures, however, two or three different coefficients for different crystallographic directions may be necessary to describe the volume change. Numerous anomalies exist, however; Invar (63% Fe, 36% Ni) expands almost not at all over a usefully wide temperature range, and $\alpha$-uranium has three widely differing expansion coefficients, one of them negative, in different crystallographic directions. The coefficient of thermal expansion for rubber is negative if the rubber is under mechanical stress but positive if the rubber is stress-free. In addition, $\alpha$ may change discontinuously at the temperature of an allotropic transformation.

The *sublimation* temperature of a material, its thermal expansion behavior, and its Young's modulus all depend on the strength of the bonds in the material. It may be seen from Figure 2.11

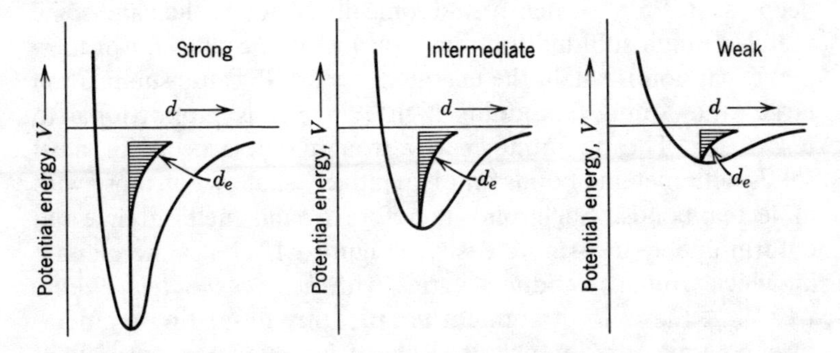


Figure 2.11 Melting point, Young's modulus, and coefficient of thermal expansion related to the shape of the Condon-Morse potential curve. The sublimation temperature is directly related to the depth of the trough; Young's modulus is inversely related to the radius of curvature at the bottom of the trough; and thermal expansion behavior is related to the degree of symmetry of the curve.

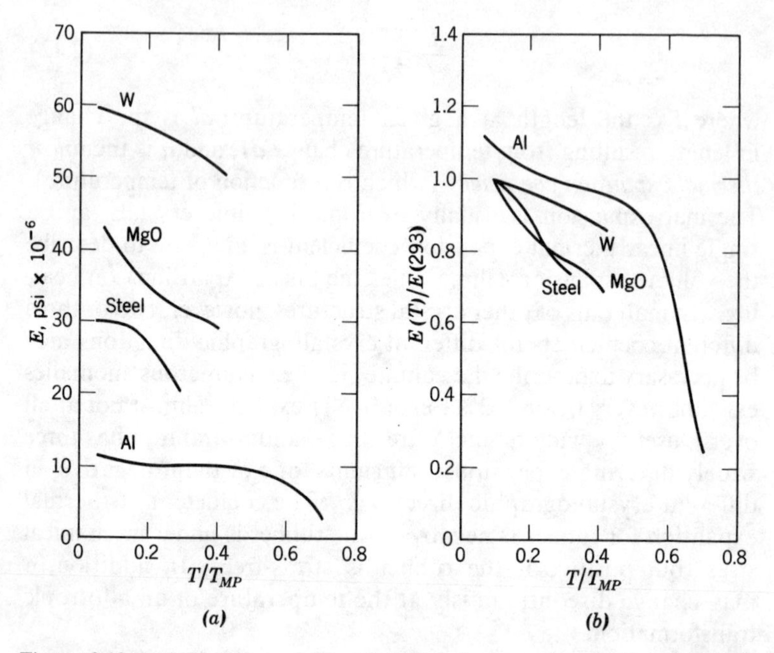


Figure 2.12 (*a*) Variation of $E$ and $T/T_{MP}$ for several crystalline materials. (*b*) Variation of $E/E_{293}$ and $T/T_{MP}$ for the same materials.

that when the Condon-Morse potential energy trough is very deep (as it is for covalently and ionically bonded solids and some metals of high sublimation temperature), the coefficient of thermal expansion is small, the energy necessary to cause sublimation large, and Young's modulus high (since it is proportional to $\partial^2 V/\partial r^2$). The potential energy trough is less deep for most solids with metallic bonds and is relatively shallow for those with molecular bonds. Such solids therefore expand, melt, sublime and deform under stress more easily. Figure 2.12 shows the manner in which Young's modulus varies with *homologous temperature* ($T/T_m \equiv$ the ratio of ambient temperature to melting temperature, both measured in degrees Kelvin) for aluminum, steel, tungsten, and MgO. A similar marked deterioration of stiffness, or lowering of Young's modulus, with increasing temperature is observed for most materials at homologous temperatures near 0.5. Static modulus measurements obtained from high-temperature stress-strain curves typically generate curves of the kind shown in Figure 2.12, but it must be noted that statically measured moduli are usually falsely low because of the complications of grain boundary relaxation and creep which affect the initial slope of the stress-strain curve. Dynamically measured moduli (i.e., moduli measured by means of a vibrating beam, for example, in which the time to reach peak stress, before stress reversal, is so short that creep addition to the elastic strain is quite negligible) are usually very much higher than statically measured moduli, at elevated temperature. Dynamic measurements indicate that the true modulus does not drop to zero as the melting point is approached.

## 2.4 EFFECT OF ALLOYING

In metals which are completely soluble in each other in the solid state, the changes in Young's modulus closely follow a mixture law across the phase diagram; in those which form eutectic mixtures with each other, Young's modulus usually shows considerable negative deviation from a mixture law (as, for example, in the individual sections $A - B$, $B - C$, and $C - D$ of Figure 2.13). In those pairs of metals, such as $A$ and $D$ in Figure 2.13, which form intermediate intermetallic compounds, Young's modulus often shows positive deviations from a mixture law (the dashed line $E_A - E_D$) with maxima at the intermetallic compounds (points $B$ and $C$) although it shows negative deviation from the mixture laws $E_A - E_B$, $E_B - E_C$ and $E_C - E_D$ between pairs of phases.

## 2.5 ELASTIC ANISOTROPY

Table 2.1 lists maximum and minimum values of Young's and shear moduli for single crystals of some common metals. The values for $E$ and $G$ of polycrystalline metals have been calculated on the basis of random grain orientation. The latter values do not hold for material which has a preferred orientation obtained in rolling or wire drawing. Such preferred orientation can be

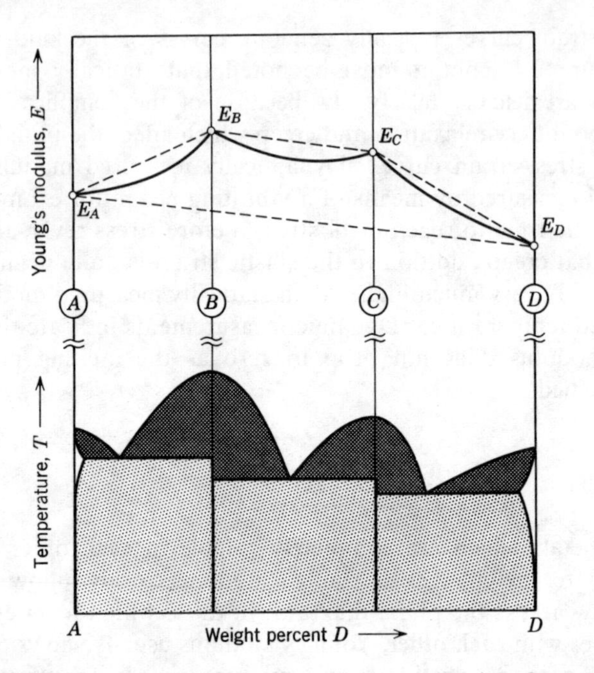


Figure 2.13 Room-temperature Young's modulus as a function of composition of binary alloys.

*Table 2.1 Young's Modulus and Shear Modulus for Some Metals*

| | E YOUNG'S MODULUS, $10^6$ PSI | | | G SHEAR MODULUS, $10^6$ PSI | | |
|---|---|---|---|---|---|---|
| METAL | MAX. | MIN. | POLY-CRYSTALLINE | MAX. | MIN. | POLY-CRYSTALLINE |
| Al | 11.0 | 9.1 | 10.0 | 4.1 | 3.5 | 3.9 |
| Cu | 27.9 | 9.7 | 16.1 | 13.9 | 4.5 | 6.6 |
| Ag | 16.7 | 6.2 | 10.4 | 6.4 | 2.8 | 4.2 |
| Pb | 5.6 | 1.6 | 2.3 | 2.1 | 0.7 | 0.90 |
| αFe | 41.2 | 19.2 | 30.0 | 16.9 | 8.7 | 12.0 |
| W | 56.5 | 56.5 | 56.5 | 22.0 | 22.0 | 22.0 |
| Mg | 7.4 | 6.3 | 6.3 | 2.6 | 2.4 | 2.5 |
| Zn | 18.0 | 5.0 | 14.5 | 7.1 | 4.0 | 5.6 |
| Cd | 11.8 | 4.1 | 7.2 | 3.6 | 2.6 | 2.8 |
| Sn | 12.4 | 3.8 | 6.6 | 2.6 | 1.5 | 2.4 |

advantageous in certain applications. It is the basis of a new development called "texture stiffening or hardening." Such texturing has long been practiced in the manufacture of silicon-iron transformer sheet for optimum magnetic properties.

## DEFINITIONS

*Elastic Strain.* A dimensional change caused by a stress which is completely recovered upon removal of that stress.
*Young's Modulus, E.* The proportionality constant between elastic strain and uniaxial stress.
*Shear Modulus, G.* The proportionality constant between elastic shear strain and shear stress.
*Bulk Modulus, K.* The proportionality constant between hydrostatic pressure and fractional decrease in volume.
*Poisson's Ratio, ν.* The ratio of transverse to axial strain caused by axial stress.
*Homogeneous.* Uniform in structure and composition.
*Isotropic.* Having uniform properties in all directions.
*Condon-Morse Curves.* Curves relating the potential energy or force of an atom or ion pair to the distance of separation.
*Linear Thermal Expansion Coefficient, α* A coefficient giving the proportionality between a linear dimensional change and the temperature change causing it.
*Homologous Temperature.* The ratio of the absolute temperature of a material to its absolute melting temperature.
*Preferred Orientation.* A nonrandom crystallographic arrangement in a polycrystalline solid.

## BIBLIOGRAPHY

### Supplementary Reading

Cottrell, A. H., *The Mechanical Properties of Matter,* John Wiley and Sons, New York, 1964, Chapters 4 and 5.
Dieter, G. E., *Mechanical Metallurgy,* McGraw-Hill Book Co., New York, 1961, Chapters 1 and 2.
McLean, D., *Mechanical Properties of Metals,* John Wiley and Sons, New York, 1962, Chapters 1 and 2.

### Advanced Reading:

Timoshenko, S., *Theory of Elasticity,* McGraw-Hill Book Co., New York, 1934.
Zener, C., *Elasticity and Anelasticity of Metals,* Chicago University Press, Chicago, 1948.

## PROBLEMS

2.1 This chapter has discussed the crystallographic anisotropy of elastic properties. Name some other types of anisotropy with examples.

2.2 (a) Name what texture cold-rolled sheet of copper, iron, and magnesium possess.

(b) Name the wire texture of W, Mo, Cu.

(c) Name the preferred texture of silicon-iron. How is it achieved.

(d) Have you ever heard of annealing texture—what is it?

2.3 How would the design engineer and the metallurgist cooperate to realize more efficient application of materials as far as modulus of elasticity is concerned?

2.4 Look up the melting points and Young's moduli of: (1) metals, (2) ionically bonded solids, and (3) covalently bonded solids.

(a) Plot Young's modulus versus absolute melting point.

(b) Briefly cxplain the pattern obtained in part a.

2.5 Define resilience.

(b) How high is it for steel (0.8 C)?

2.6 (a) Why is the resilience of strong metals increased by using them in slender shapes, such as spiral or leaf springs?

(b) In a bolt subjected to shock loads, how would a change in cross section affect matters?

2.7 How does a small circular hole disturb an otherwise uniform tensile stress in a large thin plate?

2.8 Consider a cube of homogeneous isotropic material with the stresses $\sigma_x$, $\sigma_y$, $\sigma_z$ acting on the three mutually perpendicular faces. If the cube is elastic under the influence of those three stresses, use the principle of the additive nature of elastic strains to derive the three-dimensional elastic stress-strain equations:

$$\varepsilon_x = \frac{1}{E}[\sigma_x - \nu(\sigma_y + \sigma_z)]$$

$$\varepsilon_y = \frac{1}{E}[\sigma_y - \nu(\sigma_x + \sigma_z)]$$

$$\varepsilon_z = \frac{1}{E}[\sigma_z - \nu(\sigma_x + \sigma_y)]$$

2.9 Derive Equation 2.6 for a homogeneous, isotropic material. Consider a cube subjected to a hydrostatic stress, $\sigma_{\text{Hyd}} = \sigma_x = \sigma_y = \sigma_z$ and utilize the three-dimensional stress-strain equations

$$\varepsilon_x = \frac{1}{E}[\sigma_x - \nu(\sigma_y + \sigma_z)] = \epsilon_y = \epsilon_z$$

to compute the volume change.

2.10 (a) For an isotropic material subjected to three-dimensional stressing, what relationship exists between elastic volume change and the algebraic sum of the principal stresses $(\sigma_x + \sigma_y + \sigma_z)$?

(b) What uniaxial stress $\sigma_x$ would result in the same volume change as that resulting from the three-dimensional stresses $\sigma_x = +3000$ psi, $\sigma_y = -4000$ psi, $\sigma_z = +1000$ psi?

(c) By comparison of the equation developed in (a) with Equation 2.4, what relationship exists between $\sigma_{\text{Hyd}}$ and $(\sigma_x + \sigma_y + \sigma_z)$?

2.11 Consider a cube of isotropic material stressed as illustrated:

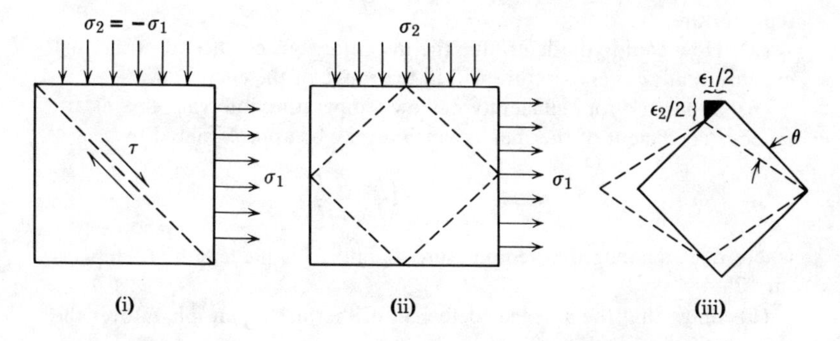


(a) Resolve forces onto the 45° plane shown and show that $|\tau| = \sigma_1$.

(b) Through use of the three-dimensional stress-strain equations given in Problem 2.8, show that $\varepsilon_2 = -\varepsilon_1$.

(c) By use of the equations

$$\varepsilon_x = \frac{1}{E}[\sigma_x - \nu(\sigma_y + \sigma_z)]$$

and $\tau = G\gamma$, show that

$$G = \frac{E}{(1+\nu)}\left[\frac{\varepsilon_1}{\gamma}\right]$$

(d) The inscribed square shown in (ii) will elastically distort into the lozenge shape indicated in (iii) in which a corresponding vertex of the undistorted square and of the lozenge are superposed. From consideration of the geometry of the small black triangle in (iii), show that $\varepsilon_1/\gamma = \frac{1}{2}$ and thus that $G = E/2(1 + \nu)$ for an isotropic material.

2.12 Calculate the ideal Poisson's ratio for a FCC lattice by assuming that the volume change associated with elastic extension is the result of stretching a (111) plane in a close-packed direction, and that the atoms act as hard spheres in contact.

2.13 A high-strength steel oil-well drill pipe, resting in its hole with its full weight on the hole bottom, is just flush with ground level at the top. The pipe is to be hoisted out for inspection of the drill bit; it is found that 16 ft of pipe are pulled out of the ground before the load in the hoisting cable stops increasing. Assume that the pipe is of uniform section along its length and that frictional effects between the pipe and the hole wall can be neglected. How deep is the hole? (Density of steel: 480 lb/ft$^3$; Young's modulus of steel: $30 \times 10^6$ psi.)

2.14 Given a curve plotting the length of a rod as a function of temperature,

(a) How could you determine the instantaneous coefficient of thermal expansion at any temperature within the range of the curve?

(b) Show that for sufficiently narrow temperature intervals, the instantaneous coefficient of thermal expansion may be approximated by

$$\alpha \simeq \frac{2}{T_2 - T_1} \cdot \left(\frac{L_2 - L_1}{L_2 + L_1}\right)$$

where $L_2$ is the length at temperature $T_2$ and $L_1$ is the length at temperature $T_1$.

(c) Show that the average coefficient of thermal expansion, $\bar{\alpha}$, over the temperature range $T_1 - T_2$ is

$$\bar{\alpha} \simeq \frac{\ln (L_2/L_1)}{T_2 - T_1}$$

(d) Show that $\bar{\alpha}$ (part c) $\rightarrow \alpha$ (part b) as $T_2 \rightarrow T_1$.

2.15 An aluminum rod, 10.000 in. long at room temperature (70°F), varies in length as a function of temperature in the following way: −400°F (9.958 in.), −300°F (9.961 in.), −150°F (9.974 in.), +200°F (10.017 in.), +400°F (10.092 in.), +950°F (10.136 in.).

(a) Plot length as a function of temperature. The curve of $\alpha$ (linear coefficient of thermal expansion) versus temperature may be obtained by differentiating the curve of length versus temperature. Observe that the slope of the length versus temperature curve changes so gradually that between two plotted points, $l_1$ and $l_2$, the slope may be taken as $\Delta l/\Delta T$. Utilize the equation

$$\alpha = \frac{1}{l}\frac{dl}{dT} \simeq \frac{2(l_2 - l_1)}{(T_2 - T_1)(l_2 + l_1)}$$

to compute $\alpha$ as a function of temperature on the same graph as the length versus temperature curve. ($\alpha$ becomes zero at the absolute zero of temperature and varies as the cube of absolute temperature at very low temperatures.)

(b) Calculate the mean coefficient of thermal expansion,

$$\bar{\alpha} = \frac{\int \alpha \, dT}{\int dT}$$

for the temperature range (room temperature to 300°F) and the range (room temperature to 800°F).

2.16 (a) Plot the coefficient of thermal expansion of copper as a function of temperature from the following data in which is given absolute temperature, °K, and, in parentheses, $\alpha \times 10^6$ in./in.°K at that temperature: 0° (0), 30° (3.0), 40° (8.0), 70° (11.4), 120° (13.2), 200° (14.6), 400° (17.0), 600° (18.3), 800° (19.7), 1000° (21.1), 1200° (22.8), 1350° (24.5).

(b) A curve of length versus temperature can be obtained by integrating the $\alpha$ versus temperature curve. Plot, on the same graph as the $\alpha$ versus temperature curve, a curve of length versus temperature choosing the reference length as unity at 0°K. Note that the area under the $\alpha$ versus temperature curve,

$$\int_{T_1}^{T_2} \alpha \, dT = \ln\left(\frac{l_2}{l_1}\right)$$

(c) Show that the choice of the reference length as unity at room temperature would not complicate the calculations.

2.17 For a homogeneous isotropic material having $\nu = \frac{1}{3}$,

(a) Compute in terms of $E$, $\Delta T$, and $\alpha$ (linear thermal expansion coefficient) the hydrostatic pressure necessary to produce at constant temperature the same fractional volume change as produced by a temperature drop $\Delta T$ acting alone.

(b) For the same material subject to a temperature increase, $\Delta T$, small enough that the elastic constants do not change in value, and to superposed stresses $\sigma_x = -6E\alpha\,\Delta T$ (compressive), $\sigma_y = -4E\alpha\,\Delta T$ (compressive), $\sigma_z = +E\alpha\,\Delta T$ (tensile), compute the resultant linear strain $\epsilon_y$.

2.18 A load $W$ is hung from three rods, two aluminum and one steel, which have the same cross-sectional area, 0.2 in.$^2$, and which were, prior to loading, the same length (at 70°F).

(a) What fraction of the total load is carried by the steel rod at 70°F?

(b) If the temperature of the whole system is changed, at what temperature will the steel rod carry all the load? At what temperature will the aluminum rods carry all the load? Take $W = 1950$ lb.

$$E_{Al} = 10 \times 10^6 \text{ psi} \qquad \alpha_{Al} = 13 \times 10^{-6} \text{ in./in.°F}$$
$$E_{steel} = 30 \times 10^6 \text{ psi} \qquad \alpha_{steel} = 6.5 \times 10^{-6} \text{ in./in.°F}$$

(c) Explain how the steel rod might carry a load greater than $W$ even though $W$ is the total weight hanging from the three rods.

2.19 One-inch cubes of aluminum and steel are set one on top of the other in a compression machine. The temperature is increased 100°F, while the pressure $p$ is also increased so that the total height of the two blocks remains exactly 2 in. Calculate, using the data for $E$ and $\alpha$ given in Problem 2.18:

(a) the pressure at the Al-steel interface.

(b) the direction of motion of the interface and the distance it moves from its initial midway position. (*Hint:* Compute the stress-free dimensions of each cube after heating and then calculate the pressure necessary to retain a 2 in. distance between the walls of the machine.)

2.20 For a gold single crystal, the measured relationship between stress and strain for normal stresses on cube faces is

$$10^8 \varepsilon_x = 16.03\sigma_x - 7.36\sigma_y - 7.362\sigma_z$$

and for shear stresses parallel to cube faces

$$10^8 \gamma_{yz} = 16.375\tau_{yz}$$

(units of the coefficients of stresses: in.²/lb). Calculate $E$, $G$, and $\nu$ for the [100] direction and compare $\nu$ with the value calculated from the isotropic relation $\nu = (E/2G) - 1$.

2.21 If $E_{100}$, $G_{100}$, $\nu_{100}$ have been determined for a single crystal of a material having a cubic structure, then in some other direction ($hkl$), having direction cosines $\alpha$, $\beta$, $\gamma$ with respect to the cube axes, $E_{hkl}$ and $G_{hkl}$ may be found from $1/E_{hkl} = 1/E_{100} - 2X$ and $1/G_{hkl} = 1/G_{100} + 4X$, where $X = [(1+\nu_{100})/E_{100} - 1/2G_{100}] \cdot (\alpha^2\beta^2 + \beta^2\gamma^2 + \gamma^2\alpha^2)$. For a silver single crystal, the measured relationship between stress and strain for normal stresses on the cube face is

$$10^8 \varepsilon_x = 15.96\sigma_x - 5.897\sigma_y - 5.897\sigma_z$$

and for shear stresses parallel to cube faces:

$$10^8 \gamma_{yz} = 15.752\tau_{yz}$$

(units of the coefficients of stresses: in.²/lb). Calculate the ratios $E_{max}/E_{min}$ and $G_{max}/G_{min}$. Is silver nearly isotropic in single crystals?

CHAPTER THREE

# Anelasticity

The migration of atoms, defects, and thermal energy are time-dependent processes. This can result in a lag of strain behind stress. The dependence of elastic strain on time as well as stress is known as the anelastic effect. In materials subjected to cyclic stress, the anelastic effect causes internal damping: a decay in amplitude of vibration and therefore a dissipation of energy. Vibrational energy in actual structures is damped out internally, in this fashion, and externally, through joint friction, wind resistance, and the like; generally the effects of the latter far outweigh those of the former mechanism.

## 3.1 INTRODUCTION

Implicit in the discussion of elastic properties in Chapter 2 was the assumption that elastic strain is a single-valued function of stress alone. This is not always the case; the attainment of maximum elastic strain can lag behind the attainment of the maximum stress causing it. The term *anelasticity* is applied to the stress-and time-dependence of elastic strain. The asymptotic approach of elastic strain to its equilibrium value with the passage of time after application of a load is known as the *elastic aftereffect.* In structures subjected to cyclic loading or to vibration, the lag of strain behind stress causes a dissipation of energy, or *damping.* Energy may also be dissipated during isothermal monotonic loading by plastic or nonrecoverable deformation. This phenomenon, known as *creep,* is treated in a later chapter. The present chapter discusses macroscopic strains which are completely recoverable, with the passage of time, after removal of the load. First we shall briefly consider several mechanisms by which the lag of strain behind stress occurs, and then we shall turn to a consideration of the experimental manifestations of the elastic aftereffect and damping.

## 3.2 THE THERMOELASTIC EFFECT

It may be demonstrated experimentally and justified theoretically that there exists an interrelationship between the mechanical work done on a material in the elastic range and changes in its thermodynamic properties, that is, between *stress* and *strain* on the one hand and *temperature* and *entropy* on the other. This relationship is known as the *thermoelastic effect.* Suppose that an elastic stress is applied to a rod so rapidly that the maximum stress is reached before the rod can exchange any thermal energy with its surroundings. The heat transferred to or from the rod is zero, so the change in internal energy is caused only by the mechanical work done on the material and the stressing is *isentropic* (that is, occurring at constant entropy and is reversible).

It may then be shown (see Problem 3.9) that for uniaxial adiabatic straining:

$$\left.\frac{\partial T}{\partial \varepsilon}\right|_S = \frac{-V_m \alpha E T}{C_V} \tag{3.1}$$

in which $\left.\frac{\partial T}{\partial \varepsilon}\right|_S$ represents the change in temperature with strain at constant entropy, $V_m$ is the molar volume of the material, $E$ is the isothermal Young's modulus, $\alpha$ is the coefficient of linear thermal expansion, $T$ is the absolute temperature, and $C_V$ is the specific heat at constant volume. Virtually all materials exhibit a volume expansion on heating. Since $\alpha$ is almost always positive, therefore, and $V$, $T$, $E$, and $C_V$ are also positive, it may be seen that adiabatic elastic tension lowers the temperature of the material and adiabatic elastic compression raises it. This temperature change is usually small, however (see Problem 3.10).

The behavior of stretched rubber provides an interesting contrast to the general behavior just discussed in that $\alpha$, the coefficient of linear thermal expansion, is negative; therefore rubber heats up on rapid stretching and cools down on rapid compression. The reason for the different sign of $\alpha$ is that, with increasing temperature, the vibration and bending of the molecular chains of which rubber is composed increase to such an extent that the mean chain length decreases, that is, the rubber shrinks. By the same token, a relatively large elastic strain imposed isothermally on a piece of rubber decreases the entropy (by straightening and tightening the molecular chains) more than it increases the internal energy.

A crystalline material tends to decrease in temperature slightly when stretched, as we have seen. If a specimen is stretched at an extremely slow rate, however, it absorbs thermal energy from its surroundings and its temperature remains constant, that is, the straining process is isothermal. Consider a specimen which exhibits a perfectly linear relationship between stress and strain under isothermal straining in its elastic region: the portion $O \rightarrow X$ of the stress-strain curve in Figure 3.1*a* and *b*. If this specimen is loaded to a stress $\sigma_1$ so slowly that the process is isothermal, the loading path is $O \rightarrow I$; if the specimen is then unloaded isothermally, the unloading path is $I \rightarrow O$. If, instead, the specimen is loaded to $\sigma_1$ so rapidly that there is no time for it to absorb thermal energy from its surroundings, the process is adiabatic and the

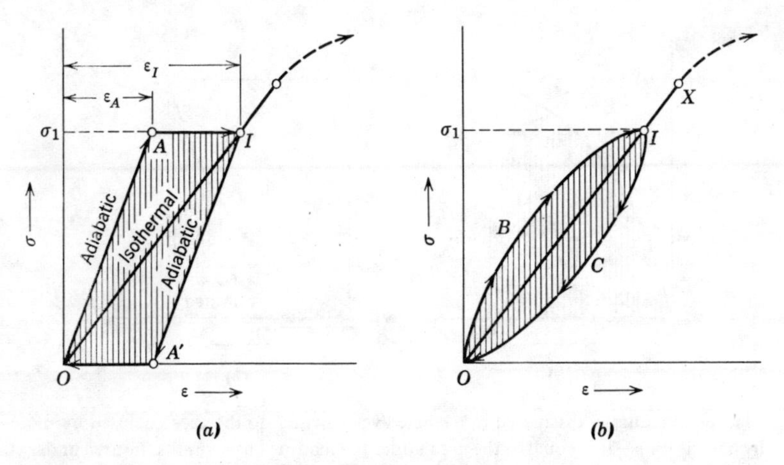


Figure 3.1 (*a*) Adiabatic and isothermal straining. (*b*) Elastic hysteresis loop.

temperature of the specimen will drop below that of its surroundings by the time $\sigma_1$ is reached. Under these circumstances the specimen follows the adiabatic loading path $O \rightarrow A$ and will accumulate only the strain $\epsilon_A$ as opposed to the larger strain, $\epsilon_I$, accumulated during isothermal stressing. If the adiabatically stressed specimen is held at the stress level $\sigma_1$, it will, with the passage of time, warm up and elongate further by thermal expansion, following the path $A \rightarrow I$. If the load is then suddenly removed, the specimen follows the adiabatic unloading path $I \rightarrow A'$ and will warm up in the process. With the passage of time the specimen transfers thermal energy to its cooler surroundings, and the strain decreases by thermal contraction following the path $A' \rightarrow O$.

The stress-strain curve of a sample loaded and unloaded in a continuous cycle would resemble the loop *OBICO* shown in Figure 3.1*b* instead of the parallelogram *OAIA'O* of Figure 3.1*a*. The shaded area in Figure 3.1*b*, called an elastic hysteresis loop, represents the energy dissipated per cycle. The elastic energy stored during the loading cycle is represented by the area under the curve *OBI* in Figure 3.2*a* (area = $\int \sigma \, d\epsilon$ and has units of lb/in.$^2$ $\times$ in./in. = in.-lb/in.$^3$, or energy per unit volume), and the elastic energy recovered during unloading is represented by

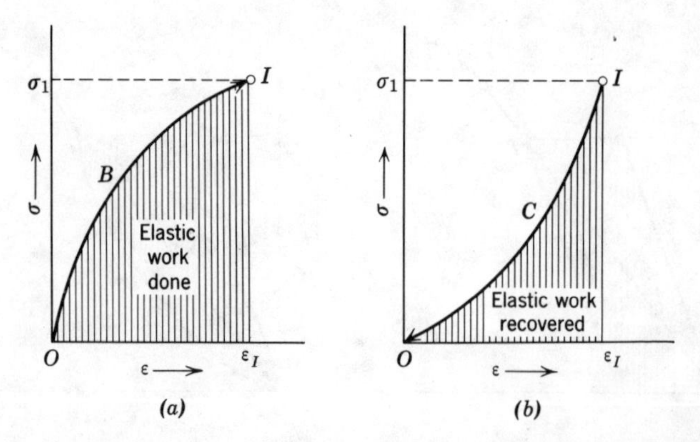


Figure 3.2 Energy dissipated in a single cycle is equal to the area enclosed by the hysteresis loop. It is equal to the area under the loading curve minus the area under the unloading curves.

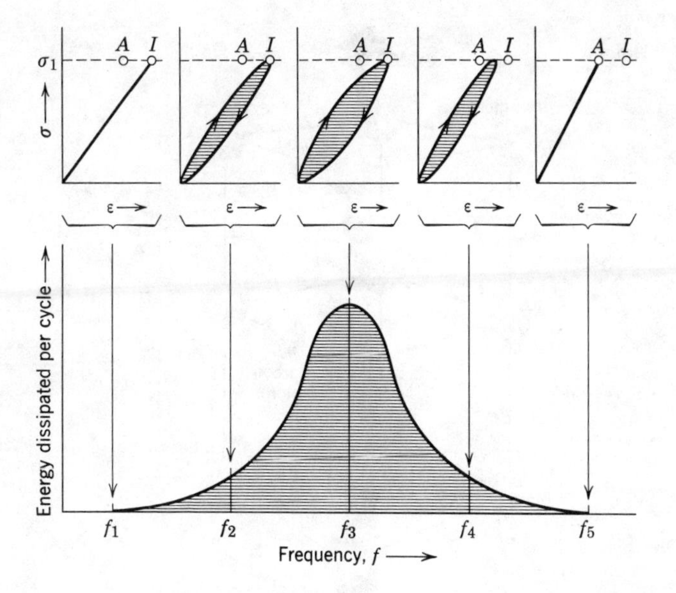


Figure 3.3 Energy dissipation per cycle as a function of frequency.

the area under the curve *ICO* in Figure 3.2*b*. The difference between the elastic work done and the elastic energy recovered is equal to the energy dissipated, which is the area enclosed by the hysteresis loop. Even though the hysteresis loop in many materials may enclose a very small area, the elastic hysteresis effect is important if the material is subject to rapid vibration, for the total energy dissipated in a given period of time is the product of the area per cycle and the number of cycles.

The area of the hysteresis loop is a function of frequency of loading and unloading: if the frequency is very low, the cycle may be almost completely isothermal, in which case the area enclosed by the hysteresis loop is extremely small. If the frequency of loading and unloading is very high, the loading and unloading paths may be almost completely adiabatic, and again the area enclosed by the hysteresis loop is very small. At some intermediate frequency, however, the area enclosed by the loop is at a maximum (see Figure 3.3).

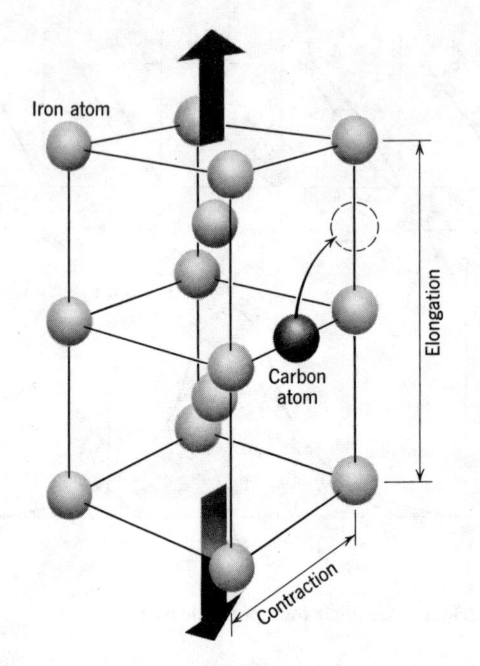


Figure 3.4 Stress-induced diffusion of carbon in iron.

## 3.3 ATOM DIFFUSION

The diffusion of interstitial atoms in a metal lattice can give rise to elastic hysteresis exactly as the diffusion of thermal energy in the thermoelastic effect. Consider, for example, the stress-induced diffusion of carbon in iron. Carbon atoms occupy random positions at the centers of edges of the body-centered cubic iron lattice, as shown in Figure 3.4. In these positions they slightly distort the unit cell. On this account only occasional carbon atoms, well separated and randomly distributed in the three perpendicular directions, can be accommodated without creating a general highly stressed state throughout the entire lattice.

Under stress, the edges of the unit cell parallel to the stress elongate and, by the Poisson effect, the edges of the unit cell perpendicular to the stress contract. Such contraction is difficult at an edge containing an interstitial carbon atom until that atom jumps into a position in an edge parallel to the applied stress (see Figure 3.4). If the stress is applied very slowly, the carbon atoms have time to diffuse to more favorable positions, and if it is released slowly, they have time to reassume a random distribution. The stress-strain relationship then resembles the isothermal loading curve of Figure 3.1*a*. On the other hand, if the specimen is rapidly loaded to a given stress and then held at that stress, the stress-strain curve resembles the adiabatic loading curve *OA* of Figure 3.1*a*. With the passage of time, carbon atoms diffuse out of unfavorable locations and into positions on the elongated edge. The removal of carbon atoms from such unfavorable sites allows a Poisson contraction, and a corresponding further elongation of the stretched cube edges, to occur. Thus, with the passage of time, the stress-strain curve is like the curve *OAI* of Figure 3.1*a*, and under cyclic loading a hysteresis loop, which is a function of frequency, is again generated.

## 3.4 GENERAL FEATURES OF ANELASTICITY

The two instances of anelastic effects discussed in Sections 3.2 and 3.3, those arising from the fact that a finite time is required for the diffusion of thermal energy or of interstitial atoms, are adequate to establish certain general features of anelasticity. Anelastic effects also arise from the motion of substitutional solute atoms, grain boundary effects, motion of dislocations, and intercrystalline and transcrystalline thermal currents. Such thermal currents have their origin in the elastic anisotropy typical of most crystalline materials. Adjacent grains strain differently and thus heat up or cool down to different extents under stress. Heat transfer then occurs from one grain to another. Whatever the mechanism by which the anelastic effect is produced, the maximum energy is dissipated when the time per cycle is of the same order as the time required for the process causing the anelastic effect. In carbon diffusion, for instance, this maximum occurs when the time per cycle is of the same order as the time required for a diffusional jump of the carbon atom. The times characteristic of the various processes giving rise to anelastic effects vary widely: for example, near room temperature the peak of the energy dissipation due to diffusion of interstitial atoms occurs at a frequency of the order of $10^{-2}$ to $10^{-1}$ cycle/sec, whereas that due to intercrystalline thermal currents occurs at about $10^5$ cycles/sec and that due to grain boundary shear, at about $10^{-8}$ cycle/sec.

The peak of a curve of energy dissipation versus frequency (Figure 3.3) occurs for a frequency at which the time per cycle is comparable to the relaxation time for the process responsible for energy dissipation. This is the average time required for the internal rearrangements which tend to occur during cyclic stressing. Diffusion rates of atoms and vacancies are highly dependent on temperature, obeying an Arrhenius-type equation as shown in Volume II. The relaxation time (as for the "jump" of a carbon atom discussed in Section 3.3) is thus also highly dependent on temperature, and it might be expected that an energy dissipation peak which occurs at a given frequency at room temperature would occur at an entirely different frequency at some other temperature. This is indeed found to be the case, and a plot of energy dissipation per cycle at constant frequency as a function of temperature has qualitatively the same form as the curve in Figure 3.3, which is a plot of energy dissipation per cycle at constant temperature, as a function of frequency. Indeed it is possible through measurement of energy dissipation as a function of temperature to identify an activation energy with the internal process causing damping. The activation energy measured in damping is compared with the activation energies measured in other ways such as in diffusion experiments. In this way certain damping peaks can be ascribed to diffusion of carbon, nitrogen, or other interstitial elements.

While a number of different internal effects contribute to the dissipation of energy, dislocation damping accounts for a large portion of the energy lost. Since a dislocation delineates the boundary between a slipped and an unslipped portion of a crystal, there is a cyclic change in the amount of slipped area as the dislocations oscillate back and forth under cyclic stress within the crystal. At low stress levels, the elastic distortion remaining in the crystal on removal of the load may be sufficient to bow the dislocations back to their original positions, and thus with the passage of time the crystal reverts to its former dimensions. In a similar fashion, during stressing of long-chain polymers, adjacent chains are locally displaced and new secondary bonds between chains are formed. This local deformation may be recovered, however, when the elastic distortion remaining after removal of the load moves the displaced segments of the chains back to their initial positions. This recovery is, of course, thermally activated, and an appreciable time may elapse before thermal fluctuations momentarily move the chain segments sufficiently far apart that the binding force between them is small enough to allow the relative motion of the two chains. Under such circumstances the hysteresis loop may enclose a very large area.

The term *damping capacity* refers to the ability of a material under cyclic stress to dissipate energy through *internal friction:* processes resulting in a loss of energy which occur within a material, such as those just described. While it is possible in carefully designed experiments to isolate the specific mechanism responsible for internal friction, in actual structures the total damping often depends much more on external factors, such as air resistance and joint friction. In addition, a number of different frequencies of vibration may be superposed, in contrast to the single frequency used in experiments, and it is usually impossible to analyze the various contributions to the total damping.

## 3.5 RELAXATION TIMES

It is desirable and often permissible to describe the time-dependent component of elastic strain with a single number—*the relaxation time,* $\tau$. To understand the meaning of this concept, consider a specimen which has a load suddenly applied at time $t = 0$ (see Figure 3.5); the specimen suffers an immediate elastic strain $\varepsilon_U$, which is the unrelaxed strain and corresponds to the adiabatic strain $\varepsilon_A$ of Figure 3.1*a*. If the load is maintained, the strain gradually increases with the passage of time toward the value $\varepsilon_R$, which is the completely relaxed strain and corresponds to the isothermal strain $\varepsilon_I$ of Figure 3.1*a*. If the load is suddenly removed at $t = t_1$, the specimen undergoes an immediate elastic contraction $\varepsilon_U$ and with the passage of time slowly approaches its initial strain-free state.

The time-dependent component of elastic strain may often be

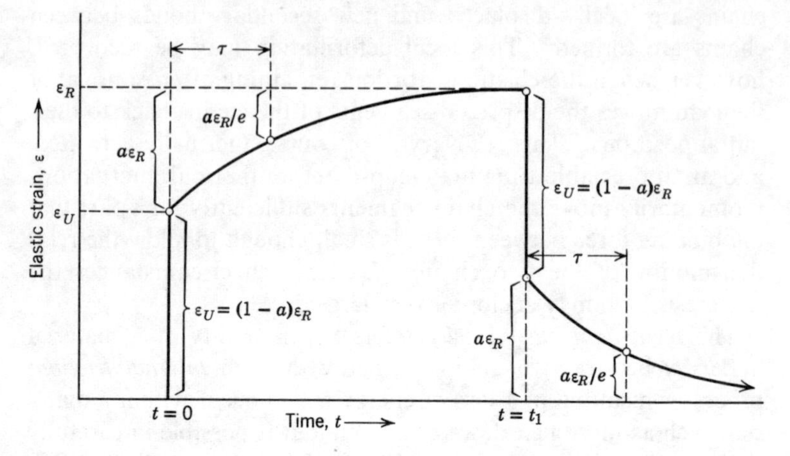


Figure 3.5 The elastic aftereffect.

approximated as an exponential function of time. If $a$ is the fraction of the total strain which lags behind application of the load, that is,

$$a = \frac{\varepsilon_R - \varepsilon_U}{\varepsilon_R} \tag{3.2}$$

the time-dependency of the loading curve may be expressed by

$$\varepsilon = \varepsilon_R[1 - ae^{-t/\tau}] \tag{3.3}$$

and that of the unloading curve by

$$\varepsilon = a\varepsilon_R e^{-[(t-t_1)/\tau]} \tag{3.4}$$

where $\tau$ is a measure of the time required for relaxation: the time required for the time-dependent component of strain to rise to within $1/e$ of its final value on loading, or to decrease to within $1/e$ of its initial value on unloading.

## 3.6 MEASUREMENTS OF DAMPING CAPACITY

An alternative to the direct measurement of relaxation time is the measurement of *damping capacity* as a function of the frequency of stressing. The measurements may be made under forced vibration or under free vibration; in the latter case, the decay of amplitude of vibration with time is measured. The specimen used is generally a vibrating beam or a torsional pendulum.

Consider a material subjected to a stress which varies as

$$\sigma = \sigma_0 \sin \omega t \tag{3.5}$$

where $\sigma_0$ is the maximum amplitude of the stress, $\omega$ is the angular frequency, and $t$ is time. Let the occurrence of maximum strain lag behind the occurrence of maximum stress by the phase angle $\phi$; then

$$\varepsilon = \varepsilon_0 \sin (\omega t - \phi) \tag{3.6}$$

where $\varepsilon_0$ is the maximum amplitude of strain. The phase difference between stress and strain, given by Equations 3.5 and 3.6, is illustrated by the elliptical stress-strain curve shown in Figure 3.6, in which it may be seen that maximum stress and maximum strain do not occur simultaneously. For example, although the stress

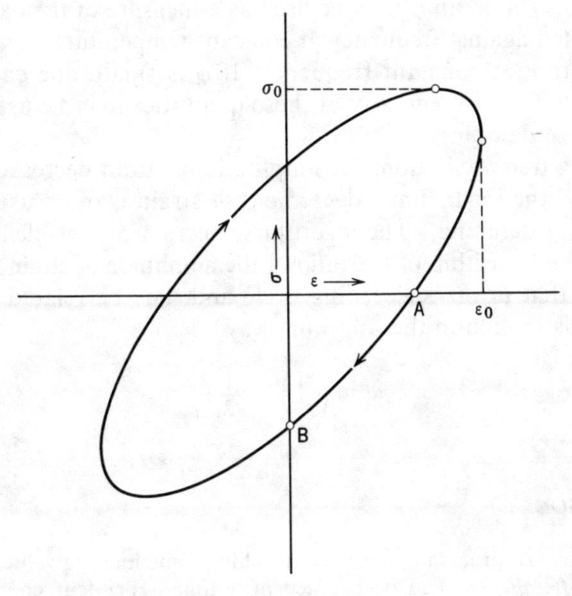


Figure 3.6 Curve showing lag of strain behind stress during cyclic loading.

has dropped to zero at point A the strain has a nonzero value; the strain drops to zero at point B, but by that time the stress has reversed sign and has a nonzero value. The loop in Figure 3.6 is a hysteresis loop for completely reversed cyclic stressing; again, the area enclosed by the loop is a measure of the energy dissipation per cycle.

The energy dissipated per cycle is

$$\Delta u = \oint \sigma \, d\varepsilon \simeq \frac{\sigma_0^2}{E} \pi \sin \phi \simeq E\varepsilon_0^2 \pi \sin \phi \tag{3.7}$$

Comparison of this energy loss with the total stored elastic energy, approximated as (see Problem 3.14):

$$u \simeq \frac{\sigma_0^2}{2E} \simeq \frac{E\varepsilon_0^2}{2} \tag{3.8}$$

yields

$$\frac{\Delta u}{u} \simeq 2\pi \sin \phi \tag{3.9}$$

Either $\Delta u/u$ or $\sin \phi$ may be used as a measure of the damping and plotted against frequency at constant temperature, or against temperature at constant frequency. If $\phi$ is small, one can take $\sin \phi \simeq \tan \phi \simeq \phi$, and any of these quantities may be used as a measure of damping.

During free oscillation, the amplitude of strain decreases with time, and the logarithmic decrement of strain is often used as a measure of damping. The logarithmic decrement, $\delta$, is defined as the natural logarithm of the ratio of the amplitude of strain in one cycle to that in the succeeding cycle and may be related to the energy dissipation in the following way:

$$\delta = \ln\left(\frac{\varepsilon_1}{\varepsilon_2}\right) \simeq \frac{1}{2}\frac{\Delta u}{u} \tag{3.10}$$

## DEFINITIONS

*Asymptotic.* Approaching but never reaching some limiting value.

*Elastic Aftereffect.* The occurrence of a time-dependent component of elastic strain.

*Damping.* The decrease in amplitude of vibration with time.
*Damping Capacity.* The ability of a material to dissipate vibrational energy, expressed mathematically as the ratio of energy dissipated per cycle to the maximum stored elastic energy.
*Internal Friction.* A general name for the mechanisms by which a material is capable of damping vibrational energy internally.
*Anelasticity.* The dependence of elastic strain on both stress and time.
*Thermoelastic Effect.* The change in temperature caused by a change in state of stress.
*Isentropic.* Occurring at constant entropy.
*Hysteresis.* The noncoincidence of the elastic loading and unloading curves under cyclic stressing.
*Relaxation Time, $\tau$.* The time required for the time-dependent component of strain to reach $1/e$ of its final value.
*Logarithmic Decrement.* The natural logarithm of the ratio of the amplitude of strain in one cycle to that in the succeeding cycle.

## BIBLIOGRAPHY

SUPPLEMENTARY READING:

Cottrell, A. H., *The Mechanical Properties of Matter,* John Wiley and Sons, New York, 1964, Chapters 4, 5.

Dieter, G. E., *Mechanical Metallurgy,* McGraw-Hill Book Co., New York, 1961, Chapters 1, 8.

McLean, D., *Mechanical Properties of Metals,* John Wiley and Sons, New York, 1962, Chapters 1, 3.

ADVANCED READING:

Brillouin L., *Wave Propagation in Periodic Structures,* McGraw-Hill Book Co., New York, 1946.

Kolsky, H., *Stress Waves in Solids,* Oxford University Press, London, 1953.

Zener, C., *Elasticity and Anelasticity of Metals,* University of Chicago Press, Chicago, 1948.

## PROBLEMS

3.1 (a) Define the term internal friction or damping capacity.

(b) What information can internal friction measurements provide?

(c) In such measurements are the vibrational amplitude and stresses high?

(d) Describe the technique according to C. Wert, *Modern Research Techniques in Physical Metallurgy,* pp. 225–250, A.S.M., Metals Park, Ohio, 1953.

3.2 (a) On what are engineering damping-capacity measurements dependent if not on frequency of vibration?

(b) Define specific damping energy.

(c) Compare the damping capacity of gray cast iron with brass, 18:8 stainless steel and cold-rolled 0.1 percent C steel.

(d) Why should the specific damping capacity at various stress levels be so very much higher for cast iron than for all other metals?

(e) Compare the damping capacity of plastics with that of brass or bronze.

3.3 How can you demonstrate that metal grain boundaries behave in a viscous manner at elevated temperatures?

3.4 Why is cast iron employed as a material for machine tool beds?

3.5 (a) Name four kinds of cast iron and compare their composition and microstructure with drawings and a table.

(b) In the same table list their elastic moduli and ductility.

3.6 (a) Lead irradiated in a nuclear reactor can be made to temporarily ring like bronze. Explain.

(b) Out of what materials are the following made and why: cymbals, church bells?

3.7 Define the word "anelasticity."

3.8 (a) When does Hooke's law appear to fail? (*Hint:* Look into elasticity of rubber and into the compression of material at great pressures as in P. W. Bridgeman, *Proc. Am. Acad. Sci.*, **72** (1938) 207; J. M. Walsh et al. *Phys. Rev.*, **97** (1955) 1544.

3.9 A free energy equation incorporating the effects of external mechanical work, analogous to that for the Helmholtz free energy, $A$, is

$$dA = -S\,dt + V_m \sigma\,d\varepsilon$$

where $S$ is entropy, $T$ is absolute temperature, $V_m$ is molar volume, $\sigma$ is a uniaxial stress, and $\varepsilon$ is the corresponding uniaxial strain.

(a) From the equality of $\partial^2 A/\partial T\,\partial\varepsilon$ and $\partial^2 A/\partial\varepsilon\partial T$ show that

$$-\left.\frac{\partial S}{\partial \varepsilon}\right|_T = V_m \left.\frac{\partial \sigma}{\partial T}\right|_\varepsilon$$

(b) Use the chain rule for partial differentials:

$$\left.\frac{\partial a}{\partial b}\right|_c = \left.\frac{\partial a}{\partial c}\right|_b \times \left.\frac{\partial c}{\partial b}\right|_a$$

to expand the result of part (a).

(c) Identify each of the terms in part (b) as $\partial S/\partial T|_\varepsilon = C_v/T$ and thus show that $\partial T/\partial\varepsilon|_S = -V_m\,\alpha\,ET/C_V$ (Equation 3.1)

3.10 From Equation 3.1: $\partial T/\partial \varepsilon|_S = -V_m \alpha ET/C_V$ compute the magnitude and sign of the temperature change resulting from adiabatic elastic uniaxial stretching of iron at room temperature (20°C) to 20,000 psi. (*Data:* Molar weight of iron: 55.85 g/mole, density of iron: 7.8 g/cc, $C_V$: approximately 6 cal/mole°K, $\alpha$: $10^{-5}$ in./in.°K, $E$: $30 \times 10^6$ psi.)

3.11 A material is subjected to periodic stressing between $\sigma_{min} = 0$, $\varepsilon_{min} = 0$ and $\sigma_{max} = 14{,}000$ psi, $\epsilon_{max} = 10^{-3}$ at a frequency for which the loading curve is given by $\sigma = K \sin 10^3\epsilon$ and the unloading curve by

$$\sigma = \sigma_{max}\left(\frac{\varepsilon}{\varepsilon_{max}}\right)\left[1 + \frac{0.1884}{2\pi}\sin(2\pi \cdot 10^3\varepsilon)\right]$$

(a) Sketch the shape of the hysteresis loop.
(b) Calculate the specific damping capacity, $\Delta u/u$.

3.12 Machine parts are seldom subjected to pure sinusoidal loading. Consider such a part subjected, by operation of a cam, to a strain:

$$\varepsilon = \varepsilon_0\left[1 + 0.9\sin\left(\frac{\pi}{4}\right)\sin\left(\frac{\omega t - (2n+1)\pi}{2}\right)\right] \qquad 2n\pi \le \omega t \le (2n+1)$$

$$\varepsilon = \varepsilon_0\left[0.3636 + 0.9\sin\left(\frac{2(n+1)\pi - \omega t}{2}\right)\right]$$

$$(2n+1)\pi \le \omega t \le 2(n+1)\pi$$

The resulting stress is

$$\sigma = \sigma_0\left[0.55 + 0.45\sin\left(\omega t - \frac{\pi}{2}\right)\right]$$

where $n$ is an integer, and $\varepsilon_0$ and $\sigma_0$ are, respectively, the maximum values of strain and stress.

(a) Sketch graphs of $\sigma/\sigma_0$ and $\varepsilon/\varepsilon_0$ and compute the specific damping capacity, $\Delta u/u$.

3.13 For a specimen subjected to a cyclic uniaxial stress such that

$$\sigma = \sigma_0 \sin \omega t$$

$$\varepsilon = \varepsilon_0 \sin(\omega t - \phi)$$

a plot of $\sigma$ versus $\varepsilon$ yields an ellipse (see Figure 3.6).

(a) For constant $\sigma_0$ and $\varepsilon_0$, sketch several of the family of ellipses corresponding to values of $\phi$ ranging from zero to $\pi/2$ and show the geometric interpretation of $\phi$ on an ellipse for $0 < \phi < \pi/2$.

(b) The area of an ellipse is $\pi ab$, where $a$ and $b$ are, respectively, the semimajor and semiminor axes. For $\phi = \pi/2$, $a = \sigma_0$ and $b = \varepsilon_0$, so the area of that hysteresis loop is $\pi\sigma_0\epsilon_0$. Show that there is no ellipse for $0 < \phi < \pi/2$ having a greater area and thus that maximum energy dissipation occurs when stress and strain are 90° out of phase.

3.14 For a specimen subjected to a cyclic uniaxial stress such that

$$\sigma = \sigma_0 \sin \omega t$$
$$\varepsilon = \varepsilon_0 \sin (\omega t - \phi)$$

(a) Show that the energy dissipated per cycle, for small $\phi$, is

$$\Delta u = \oint \sigma \, d\epsilon \simeq \frac{\sigma_0{}^2}{E} \pi \sin \phi \simeq E \varepsilon_0{}^2 \pi \sin \phi$$

(b) Evaluate the stored elastic energy, $u$, when $\sigma = \sigma_0$. Show that for small $\phi$ this expression reduces to

$$u \simeq \frac{\sigma_0{}^2}{2E} \simeq \frac{E\varepsilon_0{}^2}{2}$$

and thus $\Delta u/u \simeq 2\pi \sin \phi$ for small $\phi$.

3.15 (a) Derive an expression for the ratio of an amplitude of vibration in a given cycle to that occurring $N$ cycles later, in a freely vibrating torsional pendulum.

(b) If the frequency of vibration of a torsional pendulum is $10^{-1}$ cycles/sec and the amplitude of vibration decreases by 50 percent in 10 cycles, how long a time would be required for a 99 percent decrease in amplitude?

3.16 Both $\Delta u$ and $u$ are proportional to the square of the amplitude of vibration (see Equations 3.7 and 3.8), whereas the logarithmic decrement, $\delta$, is

$$\delta = \ln \left( \frac{A_1}{A_2} \right)$$

where $A_1$ and $A_2$ are two successive amplitudes. Show that if the angle of lag, $\phi$, is small, then

$$\delta \simeq \frac{1}{2} \frac{\Delta u}{u} \simeq \pi \sin \phi$$

Notice that if $\phi$ is small, $\Delta u$ is small compared with $u$, and since $u_1 = u_2 + \Delta u$, then $\Delta u/u \simeq \Delta u/u_1 \simeq \Delta u/u_2$. Use the expansion

$$\ln (1 + x) = x - \frac{x^2}{2} + \frac{x^3}{3} - \frac{x^4}{4} \cdots \quad -1 < x \leq 1$$

to relate $\Delta u/u$ and $\ln (A_1/A_2)$.

3.17 Young's modulus $E$ may be determined "statically" by measurement of the initial slope of a tensile stress-strain curve, or "dynamically" by several methods utilizing a vibrating beam. Under certain circumstances the two values may differ; explain how this difference could arise and illustrate the quantitative variation of $E$ with $\omega t$ on a sketch of the variation of $\Delta u/u$ versus $\omega t$.

3.18 The *modulus defect* is defined as $\Delta_0 = (E_A - E_I)/E_I$, wherein $E_A$ and $E_I$ are respectively the adiabatic and isothermal moduli associated with the adiabatic and isothermal strains, $\epsilon_A$ and $\epsilon_I$ (see Figure 3.3). $\Delta_0$ has the value $10^{-2}$ to $10^{-4}$ for many metals. Show that the maximum specific damping capacity, $\Delta u/u$, due to the thermoelastic effect is about $2\Delta_0$. Assume that the maximum energy dissipation is represented by the parallelogram shown in Figure 3.3*a*.

3.19 The relaxation time, $\tau$, for establishment of thermal equilibrium within a body, initially nonuniform in temperature, is of the order

$$\tau \simeq \frac{l^2}{D} = \frac{l^2 \rho C_p}{k}$$

where $l$ is the length of heat path, $\rho$ is density, $C_p$ is specific heat, $k$ is thermal conductivity (and $D$ is the *thermal diffusivity*). The maximum energy dissipation occurs when the relaxation time for the process responsible for damping is of the order of the time per cycle. Values of $\rho$, $C_p$, and $k$ are listed below for several elements for the 0–100°C range. Find out which element exhibits its maximum thermoelastic damping at the lowest frequency and rearrange the table in order of increasing frequency, tabulating the multiples of the lowest frequency at which maximum damping occurs in the other elements.

| | $\rho$ | $C_p$ | $k$ |
|---|---|---|---|
| Element | g/cm³ | cal/g-°C | cal/cm-sec-°C |
| Aluminum | 2.70 | 0.215 | 0.53 |
| Antimony | 6.62 | 0.049 | 0.045 |
| Bismuth | 9.8 | 0.030 | 0.02 |
| Copper | 8.96 | 0.092 | 0.94 |
| Hafnium | 13.9 | 0.035 | 0.60 |
| Iron | 7.87 | 0.107 | 0.18 |
| Lead | 11.36 | 0.031 | 0.083 |
| Molybdenum | 10.22 | 0.066 | 0.35 |
| Silver | 10.5 | 0.056 | 1.00 |
| Tungsten | 19.3 | 0.033 | 0.45 |

3.20 In a specimen vibrating as

$$\sigma = \sigma_0 \sin \omega t$$
$$\varepsilon = \varepsilon_0 \sin (\omega t - \phi)$$

the total strain, $\varepsilon$, is the sum of an adiabatic elastic strain, $\varepsilon_A = \sigma_0/E_A \sin \omega t$, and an anelastic strain $\varepsilon_a = \varepsilon_I - \varepsilon_A$.

(a) Show from Equations 3.2 and 3.3 that $\dot{\varepsilon}_a = (\varepsilon_I - \varepsilon)/\tau$ and

$$\dot{\varepsilon}_a = \left[\frac{\sigma_0}{E_I} \sin \omega t - \varepsilon_0 \sin \omega t \cos \phi + \varepsilon_0 \cos \omega t \sin \phi\right] \frac{1}{\tau}$$

(b) Evaluate $\varepsilon_a = \varepsilon - \varepsilon_A$ from $\varepsilon_A = \sigma_0/E_A \sin \omega t$ and an expansion of $\varepsilon = \varepsilon_0 \sin (\omega t - \phi)$ to show that

$$\dot{\varepsilon}_a = \left[\left(\varepsilon_0 \cos \phi - \frac{\sigma_0}{E_A}\right) \cos \omega t + \varepsilon_0 \sin \phi \sin \omega t\right] \omega$$

(c) Solve the pair of equations for $\varepsilon_a$ by equating coefficients of $\cos \omega t$ and $\sin \omega t$ in part (a) with those in part (b); eliminate $\sigma_0/\varepsilon_0$ between the resulting pair of equations to yield

$$\tan \phi = (E_A - E_I)\frac{\omega\tau}{E_A + E_I(\omega\tau)^2}$$

Since the *modulus defect,* $\Delta_0 \equiv (E_A - E_I)/E_I$ typically has a value $10^{-2} - 10^{-4}$ for most metals, $E_A \approx E_I$, and thus

$$\tan \phi \simeq \Delta_0 \frac{\omega\tau}{1 + (\omega\tau)^2}$$

(d) What is the value of $\omega\tau$ at which $\tan \phi$ has its maximum value, and what is that maximum value in terms of $\Delta_0$?

(e) Sketch a curve of $\tan \phi/\Delta_0$ versus $\omega\tau$ to illustrate the frequency dependence of $\tan \phi$.

CHAPTER FOUR

# Dislocations

Plastic, or nonrecoverable, deformation of crystalline substances occurs primarily by the movement of crystal imperfections called dislocations. The increase of stress required to produce further plastic deformation, resulting from interactions between numerous dislocations, is called work-hardening. Dislocations may exist in crystals as a result of growth faults, but in general they are produced by dislocation sources which operate under stress to disgorge dislocations successively.

## 4.1 INTRODUCTION

Application of a shear stress $\tau$ on a plane within a crystal causes a displacement, $\delta$, of atoms from their original positions, as shown in Figure 4.1. If the displacement is small, the strain is elastic; that is, on removal of the stress the atoms move back and occupy their original positions. If, however, the displacement is great enough to take atom 1 to a position midway between atoms 2 and 4, atom 1 is in a state of metastable equilibrium with respect to the two atoms and could as well take up a position over atom 4 as over its original neighbor, atom 2. It may be seen from Figure 4.1, which shows the qualitative variation of shear stress $\tau$ and potential energy $V$ with displacement $\delta$, that at the midway point no shear stress is required to cause displacement in either direction. This situation is unstable and a displacement in either direction would decrease the total energy. If under the influence of the shear stress atom 1 takes up a new position over atom 4, the symmetry of the lattice will be restored, but atoms on either side of the shear plane will have nearest neighbors different from

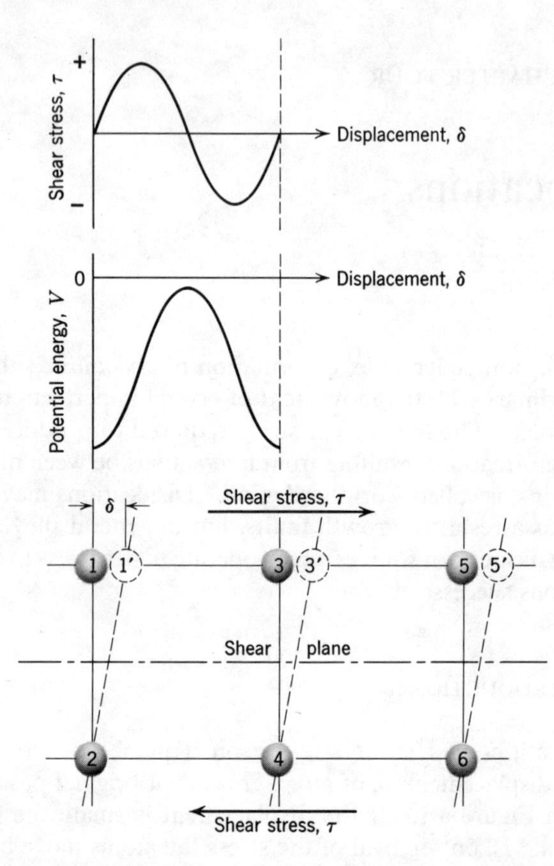


Figure 4.1 Variation of shear stress, $\tau$, and potential energy, $V$, with displacement, $\delta$.

their original ones. The crystal is then said to have *slipped,* or undergone plastic deformation.

If two complete, perfect planes of atoms are sheared over one another, the applied shear stress must overcome the attraction between each atom in one plane and its nearest neighbors on the adjacent plane. The shear stress necessary to do this has been calculated to be of the order of $1 \times 10^6$ to $2 \times 10^6$ psi; however, experimentally measured values for single FCC crystals are only of the order of $10^1$–$10^3$ psi. In order to explain the discrepancy between the theoretical and observed values of stress at which slip occurs, *dislocations* were postulated and have since been observed. The geometry of dislocations has been introduced in Volume I. The plastic behavior of crystalline materials depends on the movement and interaction of dislocations. We shall briefly consider dislocation motion before considering microscopic and macroscopic features of yielding and plastic deformation.

## 4.2 THE GEOMETRY OF DISLOCATIONS

A dislocation may be considered to have, in general, two components, an *edge* component and a *screw* component. The geometry of the lattice irregularities described by these two components, which may be treated as two simple kinds of dislocations, is shown in Figure 4.2. A dislocation is usually represented, for simplicity, by dislocation lines, such as those shown in Figure 4.2, which delineate the cores of the dislocations.

One property of a dislocation is its *Burgers* vector, **b** (see Figure 4.3), which describes both the magnitude and the direction of slip. An atom-by-atom circuit around the dislocation fails to close by

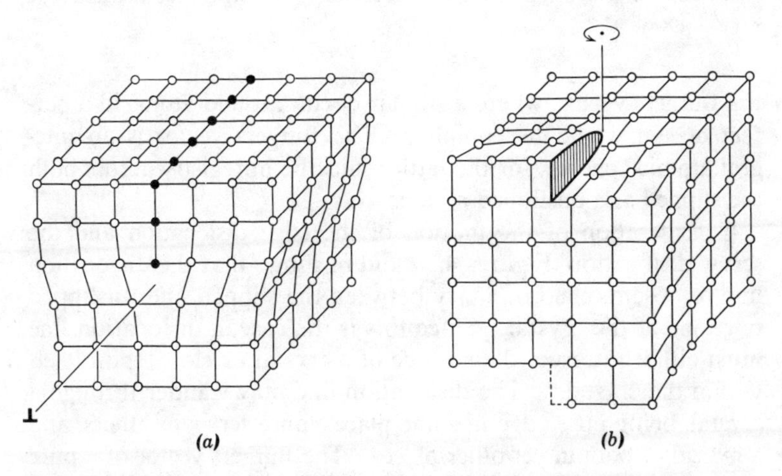


Figure 4.2 Geometry of simple dislocations. (*a*) Edge dislocation and (*b*) screw dislocation. The lines normally used to represent the dislocations are also indicated, as are the symbols for them, ⊥ and ⊙.

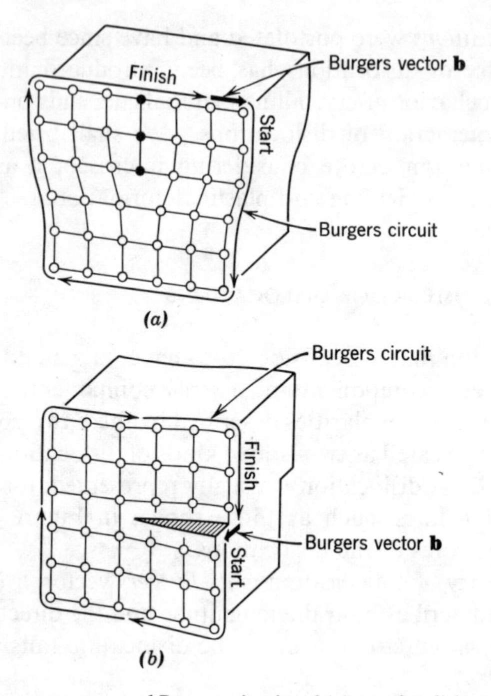


Figure 4.3 Burgers vectors and Burgers circuits of (*a*) an edge dislocation and (*b*) a screw dislocation.

the Burgers vector, where a similar circuit around atoms in a perfect crystal would be complete. The Burgers vector is an integral atomic spacing, for the lattice must be in registry across both the slipped and unslipped regions.

Consideration of the motion of the edge dislocation and the screw dislocation, Figures 4.3*a* and *b*, shows that the dislocation line represents the boundary between the slipped and unslipped portions of the crystal. Since this is the case, a dislocation line must either intersect the surface of a crystal or close upon itself within the crystal.* The dislocation line may wander through a crystal, being pure edge in some places, pure screw in others, and a hybrid of both in yet other places. The Burgers vector of a pure

* When three dislocations meet at a point (node) in a crystal, the sum of their Burgers vectors is $\mathbf{b}_1 + \mathbf{b}_2 + \mathbf{b}_3 = 0$.

edge dislocation is perpendicular to the dislocation line, that of a pure screw dislocation is parallel to the dislocation line, and the Burgers vector of a *hybrid dislocation* makes an angle with the dislocation line. As shown in Figure 4.4, a closed *dislocation loop* has pure edge character in some places and pure screw character in others, and its Burgers vector is the same at all points along the dislocation line. Note that the sign of the dislocation is taken to be opposite on opposite sides of the loop. Thus in Figure 4.4 if a clockwise circuit is made around the edge dislocation while facing the cut-out block, the Burgers vector is the same as that found by making a clockwise circuit around the screw dislocation after following the dislocation line through the crystal to the cut

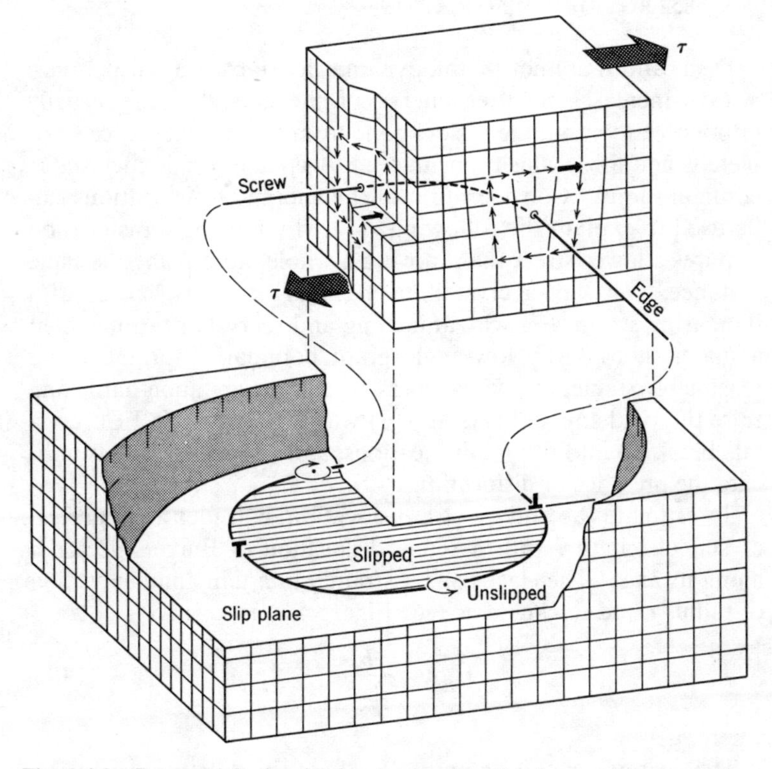


Figure 4.4 Geometry of a closed dislocation loop showing, in cut-out section, regions of pure edge and pure screw dislocation.

face from which the screw dislocation line emerges. Note that this is the same as making a *counterclockwise* circuit around the screw dislocation line while facing the cut surface from outside the block.

The Burgers vector of an edge or mixed dislocation, and the dislocation line define the slip plane. Under normal circumstances such dislocations are forced to move on this slip plane. The Burgers vector and the dislocation line of a pure screw dislocation, however, are parallel and do not define a unique plane. The screw dislocation is thus free to move on any of the several planes in which the Burgers vector lies, as discussed in Chapter 4 of Volume I.

## 4.3 ENERGY OF A DISLOCATION

Dislocations are not thermodynamically stable. Their presence always increases the free energy of the crystal. It is usually impossible to eliminate dislocations from the crystal lattice completely, and those which remain tend to assume certain metastable configurations. Consideration of the energy of dislocations can be used to explain the following: (1) why moving a dislocation requires a lower stress than moving a whole atom plane the same distance, (2) why a crystalline material becomes harder with increasing strain, (3) why annealing and recrystallization soften a material, (4) why low-angle grain boundaries form and are reasonably stable, (5) why dispersion-and precipitation-hardening raise the yield stress of crystals, (6) why dislocations in FCC crystals break up into partial dislocations, and (7) why etch pits indicate the presence of dislocations.

To estimate the energy of a dislocation, consider a cylindrical crystal of length $l$ with a screw dislocation of Burgers vector $\mathbf{b}$ along its axis. The elastic shear strain $\gamma$ in a thin annular section of radius $r$ and thickness $dr$ (see Figure 4.5) is

$$\gamma = \frac{b}{2\pi r} \tag{4.1}$$

where $b = |\mathbf{b}|$.

The energy per unit volume, $dE/dV$, of the thin annular region is then

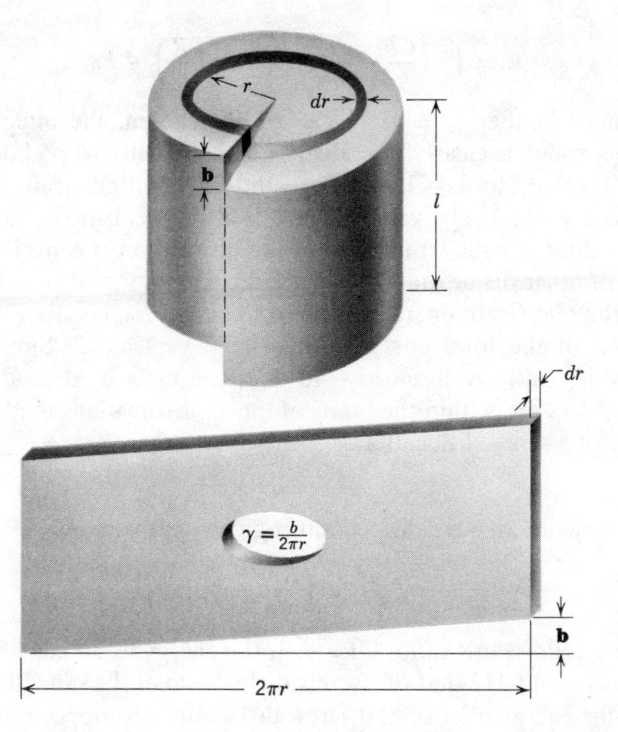


Figure 4.5 Geometric model for the calculation of shear strain around a screw dislocation.

$$\frac{dE}{dV} = \frac{1}{2}\tau\gamma = \frac{1}{2}G\gamma^2 = \frac{G}{2}\left[\frac{b}{2\pi r}\right]^2 \tag{4.2}$$

where $G$ is the elastic shear modulus. The volume of the annular ring is

$$dV = 2\pi rl\, dr \tag{4.3}$$

and thus

$$dE = \frac{lGb^2}{4\pi} \cdot \frac{dr}{r} \tag{4.4}$$

The strain energy resulting from the presence of this dislocation may be computed by integrating from some lower limit, $r_0$, to some upper limit, $R$.

$$E = \int_{r_0}^{R} l \frac{Gb^2}{4\pi} \cdot \frac{dr}{r} = \frac{lGb^2}{4\pi} \ln\left(\frac{R}{r_0}\right) + E_0 \qquad (4.5)$$

If limits of either $r_0 = 0$ or $R = \infty$ are chosen, the integral is infinite, which is clearly unrealistic. The difficulty with choosing $r_0 = 0$ is that Hooke's law is not valid for the high strain at the dislocation core. The value $R = \infty$ is also unrealistic because at large values of $r$ the strain field of the dislocation is cancelled by those of other dislocations. It has been shown that if $r_0$ is taken as $b$, the real strain energy inside of the core, $E_0$, is only a small fraction of the total energy and can be neglected. Since the energy is relatively insensitive to $R/r$, the ratio used is usually $\ln (R/r_0) = 4\pi$; within the limits of the approximations made, the energy of a screw dislocation is then

$$E \simeq lGb^2 \qquad (4.6)$$

The energy of an edge dislocation is given approximately by

$$E = \frac{1}{1-\nu} \cdot \frac{lGb^2}{4\pi} \ln\left(\frac{R}{r_0}\right) + E_0 \simeq \frac{lGb^2}{1-\nu} \qquad (4.7)$$

where $\nu$ is Poisson's ratio. If $\nu = \frac{1}{3}$, the energy of an edge dislocation is about 3/2 that of a screw dislocation of the same length. Since the energy of edge and screw dislocations is proportional to $b^2$, the most stable dislocations are those with minimum Burgers vectors (those in the close-packed directions). Equations 4.6 and 4.7 also show that the energy of a dislocation is proportional to its length; just as a surface energy is equivalent to a *surface tension,* a line energy is equivalent to a *line tension.* Thus a curved dislocation will have a "line tension", T, a vector acting along the line so that

$$T = \frac{\partial E}{\partial l} \simeq Gb^2 \qquad (4.8)$$

Figure 4.6 indicates the geometry of the *stress fields* surrounding edge and screw dislocations.

## 4.4 DISLOCATION MOTION

Consider, as the simplest possible case of dislocation motion, the edge dislocation of Figure 4.7 moving to the right in a single

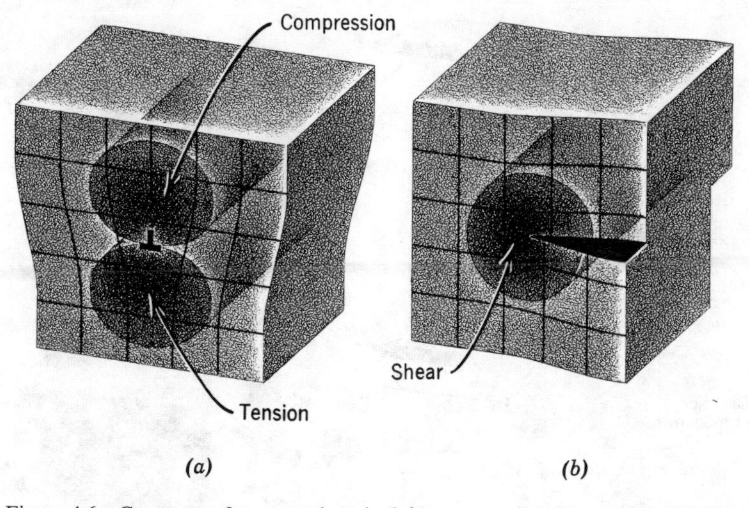


Figure 4.6 Geometry of stress and strain fields surrounding (*a*) an edge and (*b*) a screw dislocation.

crystal under the influence of the shear stress $\tau$. The shear stress must do work in pulling atom 1 further away from its neighbor, atom 2; atom 3, however, simultaneously moves closer to its equilibrium distance from its neighbor, atom 4, and relinquishes an

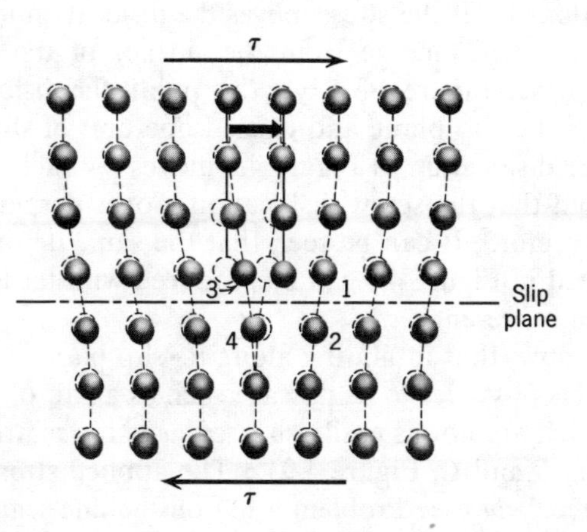


Figure 4.7 Atomic rearrangements in the vicinity of an edge dislocation as it moves under stress.

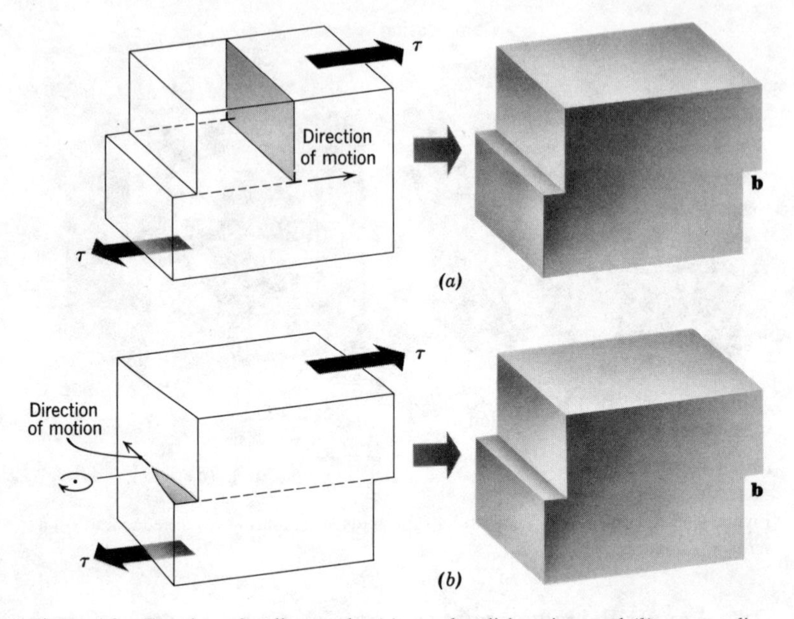


Figure 4.8 Creation of a slip step by (*a*) an edge dislocation, and (*b*) a screw dislocation, as each traverses the slip plane under the influence of the shear stress $\tau$.

amount of stored elastic strain energy almost equal to that stored in moving atom 1. If the stress moves the dislocation to the right, the dislocation will emerge from the surface of the block and create a step (see Figure 4.8). At this point, the dislocation has swept across the slip plane and caused one unit of slip. Notice that the edge dislocation in Figure 4.8 moves parallel to its Burgers vector but that the screw dislocation moves perpendicular to its Burgers vector. It can be seen that the same deformation as was illustrated in Figure 4.1 can be produced with far less work if a dislocation is present.

Suppose, now, that in gliding along its slip plane an edge dislocation meets obstacles to its passage such as a pair of precipitate particles which are not as easily sheared as is the matrix material (as at points $B$ and $C$, Figure 4.9). The applied stress $\tau$ gives a normal "force," $\tau bl$ (see Problem 4.13) on the line segment, bowing it out between the pinned points. This force is balanced by the parallel component of the dislocation line tension $T$, so that

$$\tau bl = 2T \sin\theta \tag{4.9}$$

where $l$ is the distance between $B$ and $C$. Replacement of $T$ in Equation 4.9 by its equivalent from Equation 4.8 yields

$$\tau = \frac{2Gb}{l}\sin\theta \tag{4.10}$$

It may be seen from this equation that increased stress is required to cause increased bowing of the line segment until the segment is semicircular; at this stage, $\theta = 90°$, and the stress $\tau$ assumes its maximum value:

$$\tau_{\max} = \frac{2Gb}{l} \tag{4.11}$$

According to Equation 4.10, a dislocation which does not meet any obstacles ($\sin\theta = 0$) should be capable of moving at a vanishly small stress; if it does meet obstacles, a higher stress is necessary the smaller the value of $l$. The details of what happens when the line segment becomes semicircular are considered in Section 4.8.

It is possible under certain circumstances for dislocations to avoid obstacles by changing to another slip plane. The change of slip plane by a moving screw dislocation, illustrated in Figure 4.10*a*, is called *cross-slip*. As we have seen, an edge or mixed dislocation is usually constrained to move on the plane containing

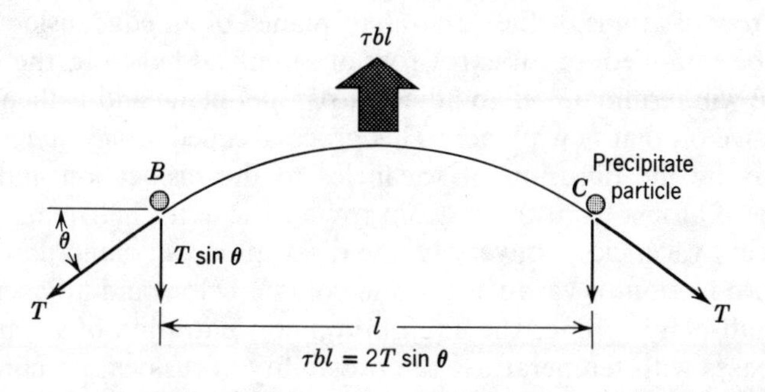


Figure 4.9 Geometric model for the calculation of the dislocation line tension brought about when the dislocation encounters the obstacles $B$ and $C$ and starts to bulge through between them.

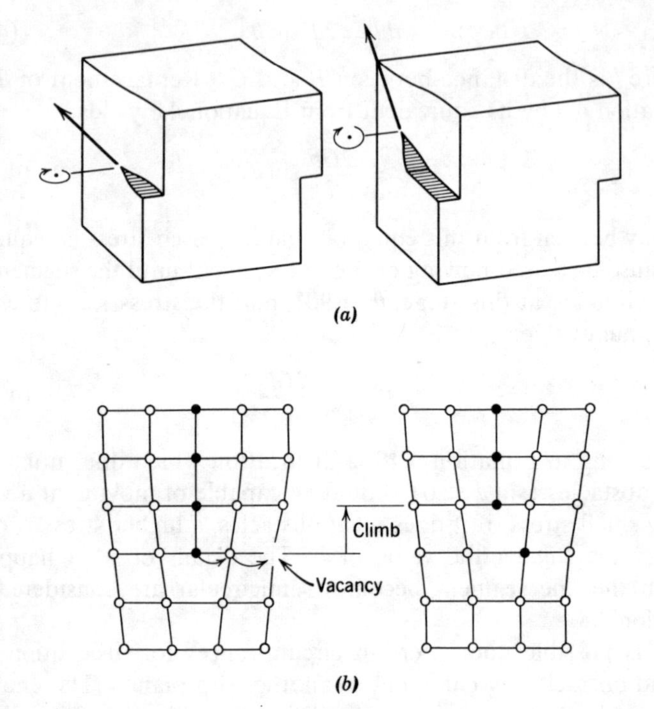


Figure 4.10 (*a*) Cross-slip of a screw dislocation, and (*b*) climb of an edge dislocation.

the dislocation line and the Burgers vector. However, if the bottom row of atoms of the "extra half-plane" of an edge dislocation can be removed, or an extra row of atoms added to it, the half-plane will terminate on an adjacent parallel plane and is then free to move on that new plane. This process, called *dislocation climb,* occurs by the diffusion of vacancies to the dislocation and the jump of atoms from the bottom row of the extra half-plane into adjacent vacancies; conversely, the dislocation can climb down by the production of vacancies in the volume below and adjacent to the half-plane. Since the equilibrium concentration of vacancies increases with temperature, the possibility of dislocation climb is temperature sensitive. Climb of an edge dislocation is illustrated in Figure 4.10*b*.

## 4.5 INTERACTIONS BETWEEN PARALLEL DISLOCATIONS

If the stress fields of two dislocations cancel, the dislocations attract each other, and if they reinforce, the dislocations repel each other. By interaction, dislocations change their total free energy. The rate of change of energy with distance gives the force between them. The radial and angular components, $(F/l)_r$ and $(F/l)_\theta$, of the force per unit length, $F/l$, between two dislocations having parallel lines are given in polar coordinates by Equations 4.12 and 4.13.

$$\left(\frac{F}{l}\right)_r = \frac{C\mathbf{b}_1 \cdot \mathbf{b}_2}{r} \qquad \text{(for both edge and screw dislocations)} \qquad (4.12)$$

$$\left(\frac{F}{l}\right)_\theta = \begin{cases} \dfrac{C\mathbf{b}_1 \cdot \mathbf{b}_2 \sin 2\theta}{r} & \text{(edge dislocations)} \\ \text{zero} & \text{(screw dislocations)} \end{cases} \qquad (4.13)$$

Here force is proportional to the *dot product* (a scalar product) of the two Burgers vectors, where $C$ is equal to $G/2\pi$ for screw dislocations and $G/[2\pi(1-\nu)]$ for edge dislocations, $\theta$ is the angle between the slip plane and the plane containing both of the dislocation lines, and $r$ is the distance of separation of the dislocation lines.

Screw dislocations of opposite sign and the same Burgers vector will attract with a force per unit length of $Gb^2/2\pi r$. If unimpeded, they come together, annihilate, and leave a perfect lattice. Screw dislocations of like sign and Burgers vector will repel with a force per unit length of the same magnitude.

A number of different interactions can occur between parallel edge dislocations because of their stress fields. Under the same shear stress, edge dislocations of opposite sign will move in opposite directions and produce slip of the same sense, as shown in Figure 4.11. Several simple interactions between parallel edge dislocations are summarized below and illustrated in Figure 4.12.

1. Two parallel edge dislocations of the same sign on the same, or nearby, slip planes repel each other. Parallel edge dislocations of the same sign on widely separated planes attract or repel each

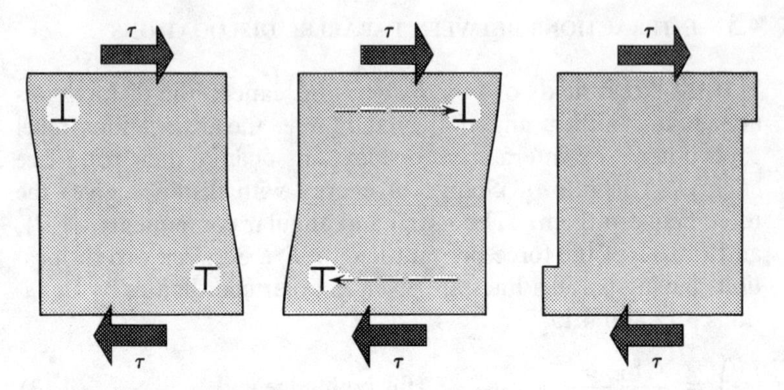


Figure 4.11 Under the shear stress, τ, dislocations of opposite sign move in opposite directions, and produce slip of the same sense.

other depending on whether the angle between the slip direction and a line joining them is greater or less than 45°, respectively.

2. Two parallel edge dislocations of opposite sign on the same, or nearby, slip planes attract each other. (a) If on the same plane, they may come together and be annihilated, leaving a perfect lattice. (b) If on adjacent planes, they may attract and be annihilated, leaving only a row of vacancies or a row of interstitial atoms.

3. Small interstitial solute atoms tend to migrate to the tension side of an edge dislocation, for there they can contribute to lowering the strain energy; small substitutional atoms, similarly, tend to migrate to the compression side of an edge dislocation, where, by replacing a larger matrix atom, they lower the strain energy. (Since no volume change is associated with a screw dislocation,

Figure 4.12 Simple interactions between parallel edge dislocations. Regions labeled *C* and *T* are, respectively, regions of compression and tension in the immediate vicinity of the dislocation.

(1) Like dislocations on same or nearby planes repel.

(2) Like dislocations on widely separated planes may attract or repel depending on the angle between slip plane and line joining the dislocations.

(3) Unlike dislocations on same or nearby planes attract.

  (*a*) If they are on the same plane, they annihilate and leave perfect lattice.

  (*b*) If on adjacent planes, they annihilate and leave vacancies or interstitials.

(4) Small interstitial atoms are attracted to tension side of dislocations.

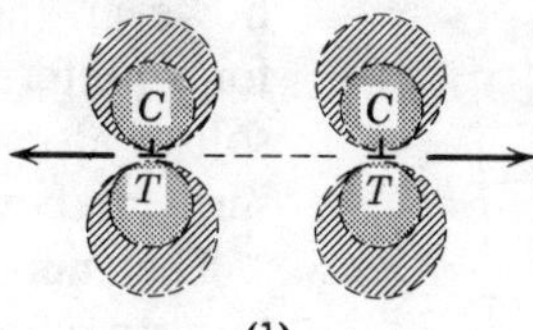
C
C
T
T

(1)

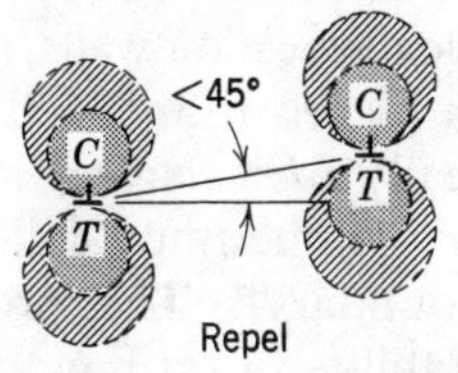
<45°
C
C
T
T
Repel

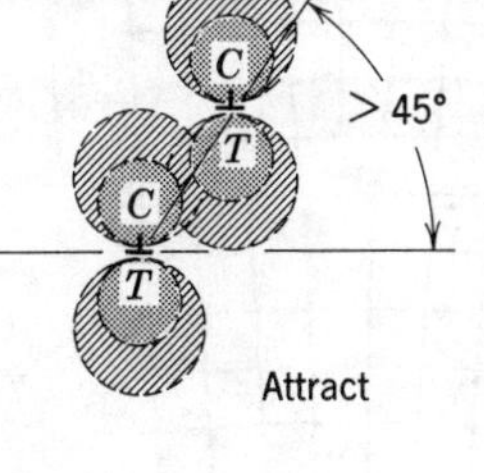
C
T
C
T
>45°
Attract

(2)

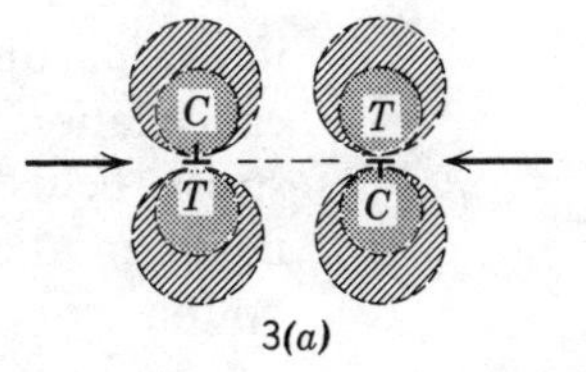
C
T
T
C

3(a)

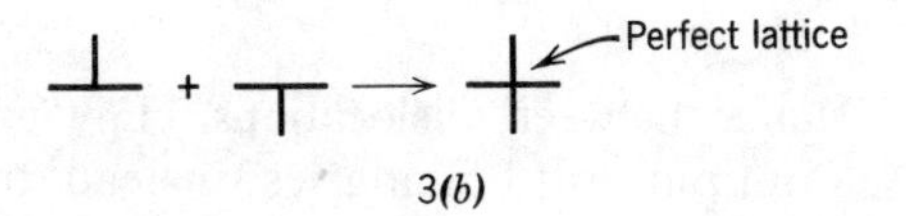
Perfect lattice

3(b)

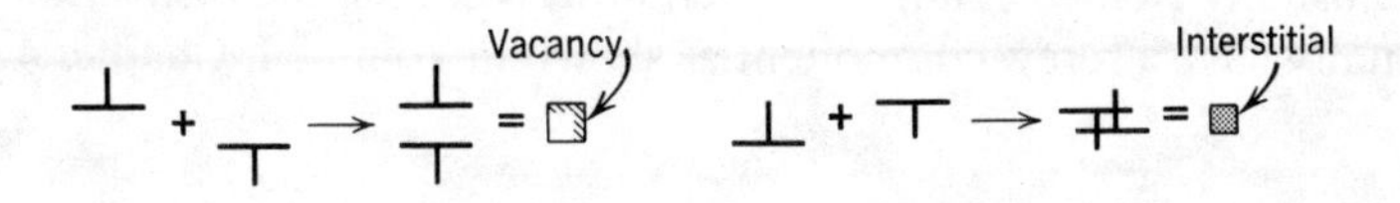
Vacancy
Interstitial

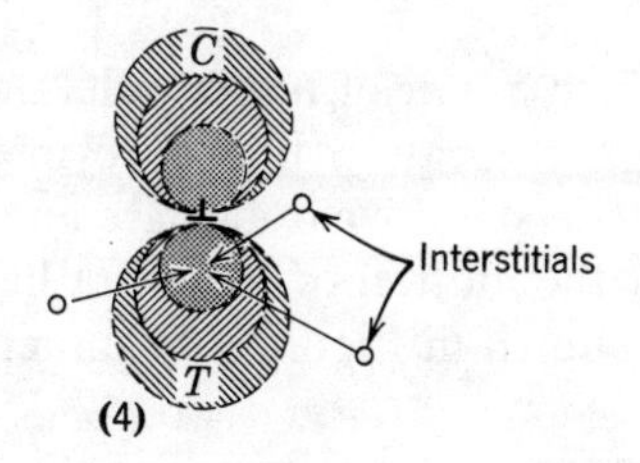
C
Interstitials
T

(4)

however, there is no tendency for a preferential relocation of solute atoms of the kind described above.)

The energy of a crystal containing edge dislocations can be reduced if the edge dislocations line up one above the other, producing relatively stable dislocation walls, as shown in Figure 4.13. Such a wall is really a *low-angle grain boundary,* sometimes called a *tilt* boundary. The provisional stability of tilt boundaries results from the absence of stress on the slip planes of the individual dislocations and the cancellation of long range stress fields. The angle of misorientation across such a boundary is

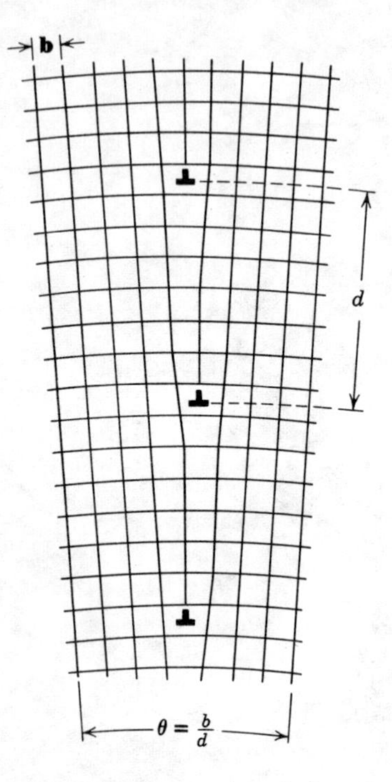


Figure 4.13 Low-angle grain boundary produced by the alignment of edge dislocations of the same sense.

$$\sin\theta \simeq \theta \simeq \frac{b}{d} \quad (4.14)$$

where $d$ is the distance between dislocations. Low-angle grain boundaries are seldom pure tilt boundaries; instead, the lattices are usually twisted relative to each other (that is, the dislocations usually have a screw component as shown in Volume I, Chapter 4).

## 4.6 INTERACTION BETWEEN INTERSECTING DISLOCATIONS

When a pair of nonparallel dislocations intersect, a jog or step in the dislocation line is formed in either or both dislocations. To understand what jogs are and to see how they are formed, consider the motion of the extra half-plane $DD$ of the edge dislocation shown in Figure 4.14 under the influence of a shear stress $\tau$.

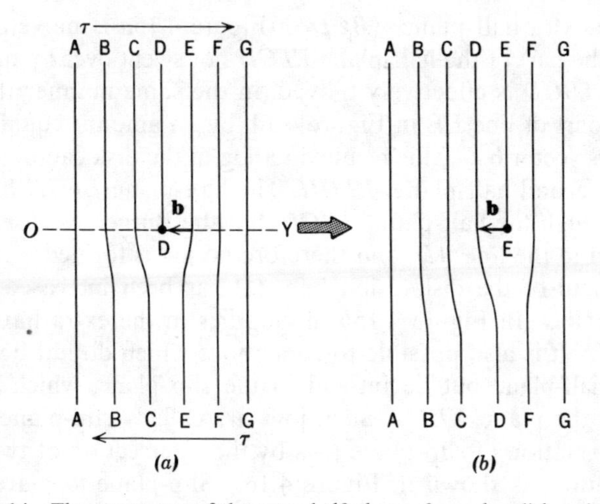


Figure 4.14 The movement of the extra half-plane of an edge dislocation under an applied stress.

As the line of the extra half-plane is moved along its slip plane in the slip direction $OY$, the bottom half of the row of atoms EE is shifted one Burgers vector. In Figure 4.15 the intersection of two edge dislocations with extra half-planes $ABCD$ and $EFGH$ is shown. The Burgers vectors are $\mathbf{b}_A$ and $\mathbf{b}_E$, as shown in the figure.

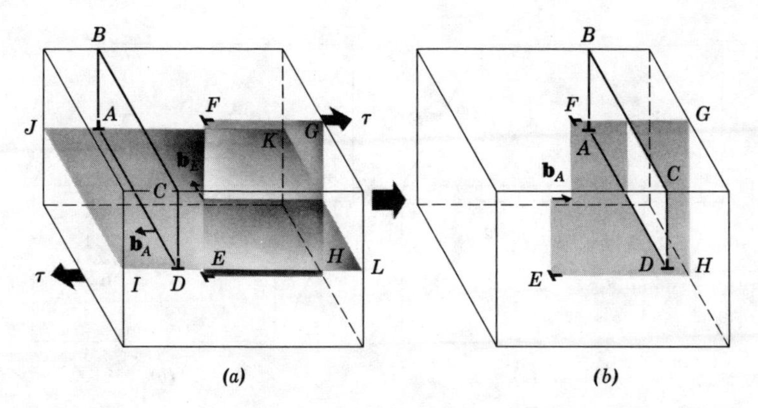


Figure 4.15 An example of the intersection of two edge dislocations to create a jog.

As the extra half-plane *ABCD* in Figure 4.15*a* is moved to the right, the part of the half-plane *EFGH* not swept over by the half-plane *ABCD* is effectively moved, in the same manner that the lower part of line EE in Figure 4.14, by an amount equal to the Burgers vector $\mathbf{b}_A$. This results in a jog in the dislocation line *EF* of the second half-plane, *EFGH*. The half-plane *EFGH* has also cut through the half-plane *ABCD*, but the Burgers vector $\mathbf{b}_E$ lies parallel to the line *AD*, and therefore no jog is formed. Instead, the length of the dislocation line *AD* has been increased by an amount $\mathbf{b}_E$. In Figure 4.15*b* the jog lies in the extra half-plane *EFGH*. It is also possible to create jogs which do not lie in the extra half-plane but lie instead in the slip plane, which in this case is the plane *KLIJ*. Such jogs are called "slip-plane jogs." The formation of slip-plane jogs by the intersection of two edge dislocations is shown in Figure 4.16. Slip-plane jogs are more readily eliminated than half-plane jogs. In the former case one part of the dislocation half-plane can easily move in the slip plane back into alignment with the other part, whereas in the latter case it is necessary for the dislocation line to *climb* in order for the jog to be eliminated.

Intersection between an edge and a screw dislocation or two screw dislocations is also possible. In general, when two dislocations intersect, jogs are created that are equal in length to the component of each Burgers vector normal to the other dislocation

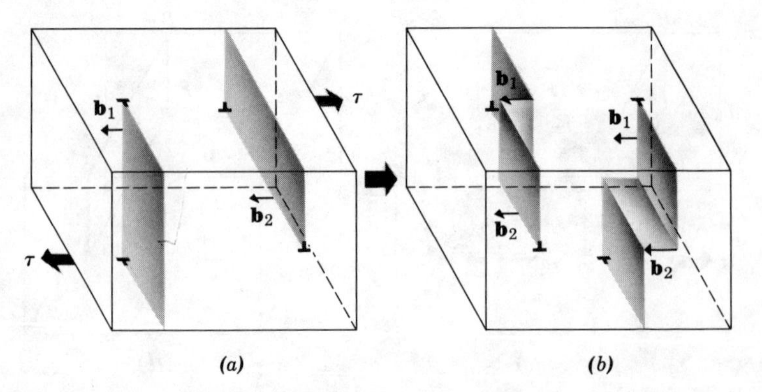


Figure 4.16 An example of the intersection of two edge dislocations to create two "slip-plane jogs."

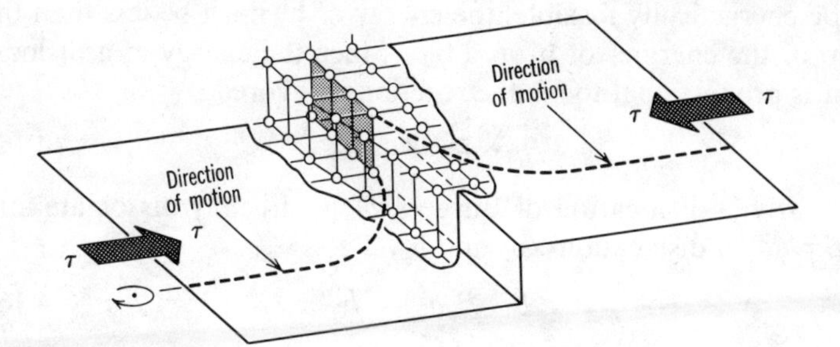


Figure 4.17 Motion of a jogged screw dislocation leaves behind a row of vacant lattice sites.

line. The Burgers vector of the jogged section is always the same as that of the dislocation line of which it is a part. Thus the slip-plane jogs in Figure 4.16 are short sections of screw dislocations, for the line of the jog is parallel to the Burgers vector.

The presence of jogs makes the dislocations containing them less mobile. As an example, consider the jogged screw dislocation of Figure 4.17; while it is, for the most part, a screw dislocation (the dislocation line lies parallel to **b**), the jog is a segment of edge dislocation (it lies perpendicular to **b**). The edge dislocation (the jog) cannot move to the right without diffusion of atoms to it to extend the half-plane. When the screw part of the dislocation moves, a double trail of dislocations (which is equivalent to a row of vacant lattice sites) is left behind. Similar considerations apply to the hindrance of the motion of jogged edge dislocations. A greater amount of work is required to move a jogged dislocation than a straight one. The successive intersection of dislocations during plastic deformation makes an ever-increasing stress necessary in order to cause further dislocation motion; this is probably a very important contributing factor in the strain-hardening of crystals.

## 4.7 DISLOCATION REACTIONS

Two parallel dislocations of Burgers vectors $\mathbf{b}_1$ and $\mathbf{b}_2$ can react to form a third dislocation of Burgers vector $\mathbf{b}_3$. For the reaction to be energetically feasible, the energy of $\mathbf{b}_3$ must be less than the sum of the energies of $\mathbf{b}_1$ and $\mathbf{b}_2$. Since the energy of a dislocation is proportional to $b^2$, the reaction is favored if

$$b_1^2 + b_2^2 > b_3^2 \tag{4.15}$$

Similarly, a dislocation of Burgers vector $\mathbf{b}_1$ may dissociate into two parallel dislocations, $\mathbf{b}_2$ and $\mathbf{b}_3$, if

$$b_1^2 > b_2^2 + b_3^2 \tag{4.16}$$

Both the crystallographic direction and the magnitude of a Burgers vector are related to the distance between atom sites. In a cubic crystal, the direction can be expressed by the Miller indices, $[hkl]$, and the magnitude by $ca\sqrt{h^2 + k^2 + l^2}$, where $c$ is some fraction and $a$ is the lattice parameter. Often the simple notation $ca[hkl]$—for example, $a/2[011]$—is used to describe a Burgers vector whose direction is $[hkl]$ and whose length is $ca\sqrt{h^2 + k^2 + l^2}$. Therefore

$$b^2 = c^2a^2(h^2 + k^2 + l^2) \tag{4.17}$$

In FCC crystals, $\langle 110 \rangle$ slip occurs by a zig-zag path corresponding to $\langle 21\bar{1} \rangle$ plus $\langle 121 \rangle$ displacements on $\{111\}$ planes, as shown in Figure 4.18*a*. If a $\langle 21\bar{1} \rangle$ zig occurs without a $\langle 121 \rangle$ zag, the atoms in the slip plane are directly over the atoms of a plane two planes beneath it, and the local stacking is that of the HCP structure. This would occur if a $\langle 110 \rangle$ dislocation dissociated into partial dislocations (that is, dislocations which do not alone complete a unit slip displacement) corresponding to the $\langle 21\bar{1} \rangle$ zig and the $\langle 121 \rangle$ zag. The value of $b^2$ for the $\langle 110 \rangle$ dislocation is greater than the sum of $b^2$ for the two partials, and thus this dissociation is energetically favorable. The partial dislocations repel one another and tend to separate and create a *stacking fault*—the local region of HCP structure (Figure 4.18*b*). In a metal which is normally FCC, the local HCP configuration has a higher energy (otherwise the metal would be normally HCP), so the faulted region is one of relatively high energy. The energy of the crystal would be reduced if the area of the faulted region were decreased, but the partial dislocations, because they both have edge character of the same sign and are on the same slip planes, repel each other. The forces of attraction and repulsion balance at an equi-

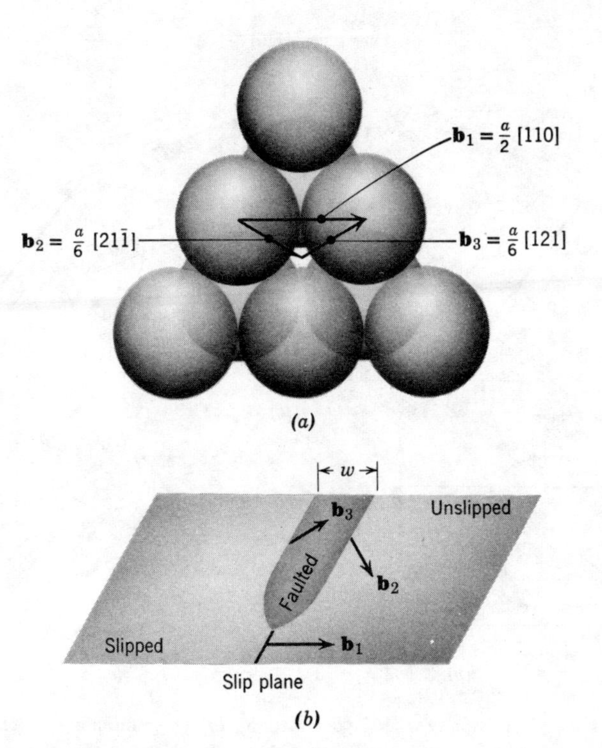


Figure 4.18 (*a*) ⟨110⟩ slip in FCC crystals. The unit slip vector of the complete dislocation, $\mathbf{b}_1$, is $a/2$ [110]; two partial slip vectors, the sum of which produces the same net motion as $\mathbf{b}_1$, are $\mathbf{b}_2$ and $\mathbf{b}_3$. (*b*) Faulted region of width *w* between the two partial dislocations $\mathbf{b}_2$ and $\mathbf{b}_3$.

librium spacing that depends on the specific stacking-fault energy. Thus dislocations in FCC metals are normally dissociated into two partial dislocations connected by a stacking fault.

While whole screw dislocations are free to move (and cause slip) on any of several planes in which they lie, dissociated screw dislocations, since they have edge character, can move only on the plane containing the stacking fault. If a screw dislocation is to change slip planes, its partials must first be recombined; it can then dissociate into a different pair of partial dislocations on a new slip plane. Both stress and thermal fluctuations affect the critical recombination step. At normal temperatures, the recom-

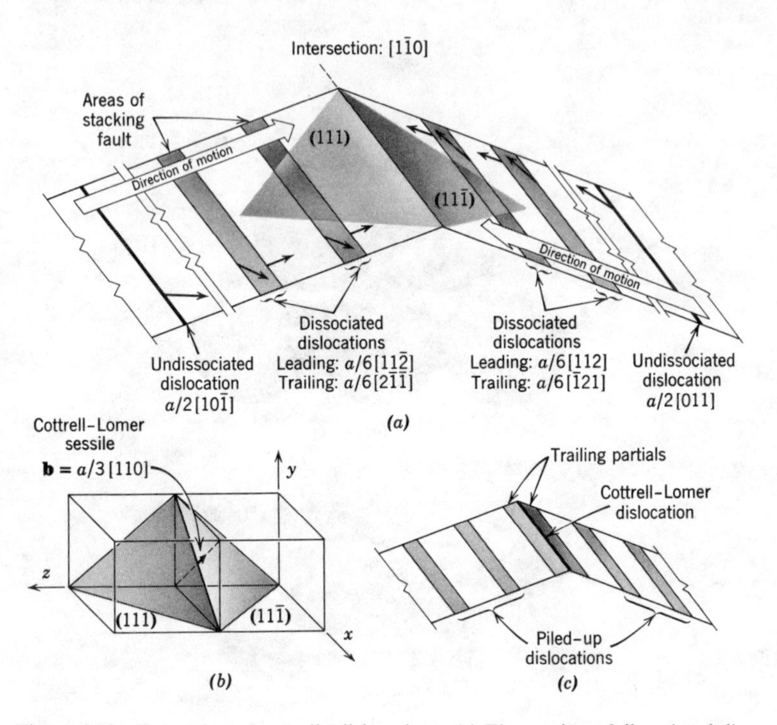


Figure 4.19 Formation of a sessile dislocation. (*a*) The motion of dissociated dislocations on two intersecting slip planes, (111) and (11$\bar{1}$); (*b*) the relative orientation of the two intersecting planes in (*a*), and (*c*) the combination of leading partial dislocations moving on the two intersecting planes to produce a sessile dislocation whose Burgers vector lies in neither plane—see (*b*).

bination process in FCC metals is sufficiently difficult that slip traces which change planes are usually observed only at high stresses. In BCC metals, however, dislocations apparently cannot dissociate easily. The necessity of recombination of partials is therefore absent, and wavy slip traces caused by cross-slip are commonly observed.

Two parallel dislocations on different planes may meet and react to form a *sessile*, or immobile, dislocation. Such dislocations cannot move by the slip process alone because the Burgers vector is not contained in the slip plane. One such dislocation, called a *Lomer-Cottrell* dislocation, is shown in Figure 4.19. The particular sessile dislocation shown is that produced by two dislocations moving on a pair of $\{111\}$ planes, the active slip planes in the FCC lattice; sessile dislocations may, however, be produced by dislocation reactions other than this. Figure 4.19*b* shows the orientation of the two planes with respect to the crystal lattice; these planes are also indicated in Figure 4.19*a* by the two triangular regions.

Consider a pair of parallel dislocations, $a/2[011]$ and $a/2[101]$, each dissociated into partial dislocations, as shown in Table 4.1.

*Table 4.1*

| Complete Dislocation | | Partial Dislocations | | |
|---|---|---|---|---|
| | | Leading | | Trailing |
| $\frac{a}{2}[011]$ | $\longrightarrow$ | $\frac{a}{6}[112]$ | + | $\frac{a}{6}[\bar{1}21]$ |
| $\frac{a}{2}[10\bar{1}]$ | $\longrightarrow$ | $\frac{a}{6}[11\bar{2}]$ | + | $\frac{a}{6}[2\bar{1}\bar{1}]$ |

The leading partial dislocations may then meet and react to produce a new dislocation:

$$\frac{a}{6}[112] + \frac{a}{6}[11\bar{2}] \rightarrow \frac{a}{3}[110]$$

Comparison of the sum of $b^2$ for the two partials with $b^2$ for the new partial dislocation shows that the latter is energetically favored. However, the new dislocation lies along a face diagonal of the unit cell (Figure 4.19*b*), and its Burgers vector, $a/3[110]$, also lies in the cube face along a diagonal perpendicular to the dislocation line. The new dislocation thus has edge character and can move only on the cube face; this, however, is not an active slip plane at normal temperatures in FCC crystals. The consequence of this reaction, as shown in Figure 4.19*c*, is that the trailing partials and following dislocations pile up behind the Lomer-Cottrell sessile dislocation, and an ever-greater force is necessary to push additional dislocations into the pile-up.

The condensation of vacancies can also create sessile dislocations. Because the equilibrium solubility of vacancies decreases with decreasing temperature, a crystal lattice may become super-

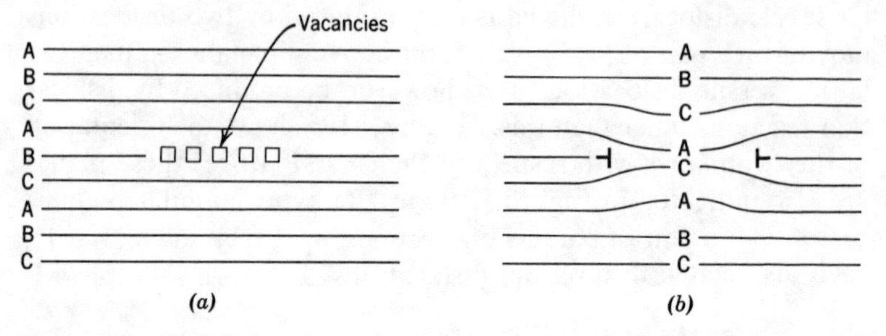


Figure 4.20 (*a*) Collapse of vacancy cluster to form (*b*) dislocation loop enclosing an area across which the stacking order differs from that characteristic of the crystal.

saturated with them during cooling. If the cooling rate is too rapid for the vacancies to move to sinks, such as surfaces, grain boundaries, or dislocations, they may precipitate into clusters which, when large enough, collapse into prismatic dislocation loops (see Figure 4.20). These loops are sessile in FCC materials, where they enclose an area of stacking fault perpendicular to the plane in which the loop lies.

## 4.8 WORK-HARDENING AND RECRYSTALLIZATION

A very low stress is required to move a single, isolated dislocation through a crystal. The strain produced when it runs out of the crystal is also small, and only a single slip step of magnitude $b$ is produced. Clearly, then, the generation of macroscopic strain requires the cooperative motion of a very large number of dislocations.

Let the density of dislocations, $\rho$, in a crystal be expressed as the length of dislocation line per unit volume—cm of line/(cm)$^3$—or, what is approximately but not precisely the same, by the number of dislocation lines piercing a unit area of a random section—lines/(cm)$^2$. Very pure well-annealed single crystals may have dislocation densities as low as $10^2$–$10^3$ lines/cm$^2$. If all these dislocations were arranged to act cooperatively and all were able to run out of the crystal with ease, then the total shear strain produced in crystals 1 cm in diameter would be of the order of $10^{-5}$, well below the shear strain which corresponds to an 0.2 percent offset yield strain. How can we explain the fact that such a crystal can often be extended to very large strains, and, indeed, that increasing stress is generally necessary to produce further strain?

The initial dislocation density in a metal is usually many orders of magnitude lower than that required to produce the eventual fracture strain. For example, a typical dislocation density for annealed polycrystalline metals is about $10^7$–$10^8$ lines/cm$^2$, whereas the same specimen, after severe plastic deformation, contains $10^{11}$–$10^{12}$ lines/cm$^2$. Some mechanism, therefore, must exist by which dislocations can be produced during plastic deformation.

One such dislocation source, a *Frank-Read source,* is illustrated in Figure 4.21. Figure 4.21*a* shows, for reference, a crystal containing an edge dislocation with a *half-plane jog* and Figure 4.21*b* shows only the dislocation line. The dislocation need not be the jogged edge dislocation shown for a Frank-Read source to operate in the way described below but could be any mixed dislocation containing the same jog.

Under the influence of the applied shear stress $\tau$, the line segment BC starts to move; the segments AB and CD do not, however, for there is no resolved shear stress on their slip planes. Segment BC cannot move very far, however, before starting to drag the other two segments with it. If the dislocation line is pinned at points B and C, the segment BC bows out until the force $\tau_{bl}$ is balanced by the increase in line tension resulting from the increase in length of the dislocation line. If the shear stress is sufficiently high, the dislocation expands and passes around the pinning points, B and C. At this stage, screw dislocations of opposite sign exist at points 1 and 2; these attract, annihilate each other, and leave a region of perfect lattice. The remaining parts of the line, being edge dislocations of the same sense, can lower their energy by joining together. Thus the original dislocation forms a complete loop, and the initial line segment *BC* is reformed; the line segment can then repeat this process and create further dislocation loops. The completed loop is the boundary between the slipped and unslipped part of the crystal; as the loop expands

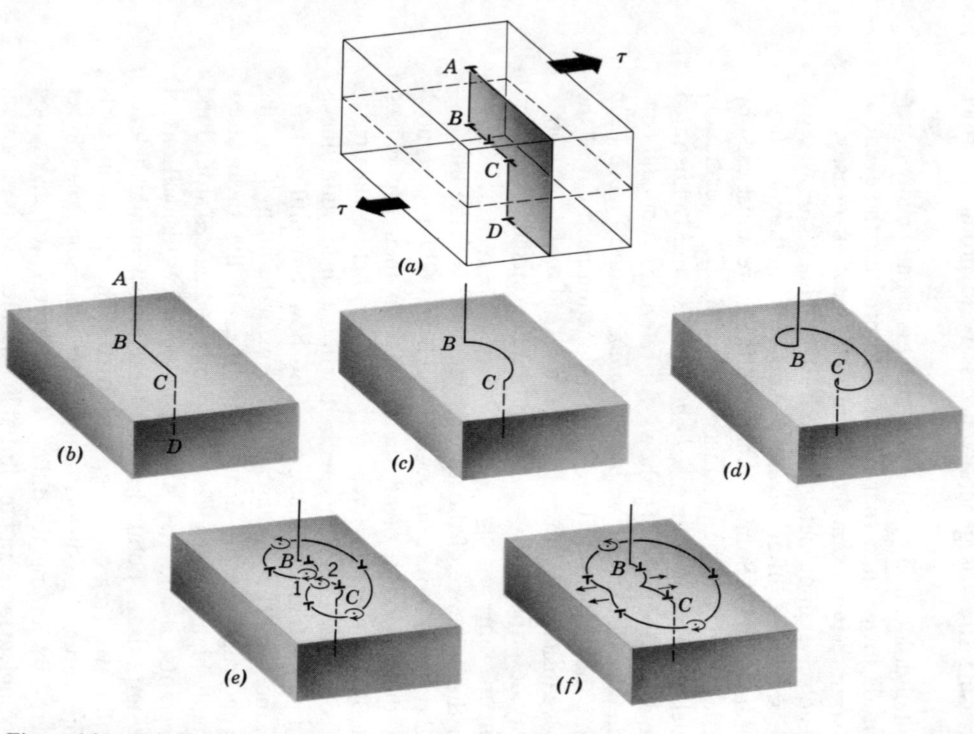


Figure 4.21 The Frank-Read dislocation source. (*a*) A crystal containing an edge dislocation with a half-plane jog; (*b*) the line representing the dislocation core. The dislocation is pinned at points *B* and *C*. The subsequent parts of the illustration show the sequence of events as the dislocation line bulges through between the pinning points, swings around them, and meets on the far side to make a complete dislocation loop and reform the initial line segment *BC*.

and passes out of the crystal, a single unit of slip will occur, creating a slip step at the right and left surfaces.

As may be seen from Equation 4.11, the largest sources (i.e., those with the greatest distance, $l$, between the points at which the dislocation is pinned) operate at the lowest stresses; if the loops emanating from this source begin to pile up at some obstacle, a greater stress is required to continue the deformation. This increased stress will then be large enough to allow smaller sources to operate.

The operation of numerous sources of this type causes a continual increase in the frequency of dislocation intersections and, eventually, in a *tangled forest* of dislocation lines, all containing numerous jogs. Thus, ever-increasing stress is required to cause further strain. This mechanism is called strain-hardening. The *rate* of stress increase, however, usually diminishes with increasing strain, for there are recovery, or softening, processes which work in opposition to the strain-hardening mechanism. For example, although the applied stress must increase in order to create further strain, this same increase in stress may also activate cross-slip and force piled-up dislocations to move. With increased temperature, the recovery process of dislocation climb may also be activated.

In many commercial metal-working processes, there is a limit to the amount of strain to which a part may be subjected without danger of cracking or tearing. For this reason the material is often *annealed,* or heated above its recrystallization temperature for a predetermined period of time. New, strain-free grains are produced by this treatment. The softened material is then capable of further deformation. Annealing and recrystallization account for the elimination of a large number of dislocations and the rearrangement of others into lower-energy configurations. Many dislocations run out of the surface of the material, disappear at sinks, such as grain boundaries or microscopic voids, or are annihilated by other dislocations of opposite sign. Those which cannot do any of the foregoing tend to change their configuration; jogs may disappear and dislocation lines shorten, or similar dislocations may be rearranged into subgrain boundaries, such as the low-angle grain boundary shown in Figure 4.13.

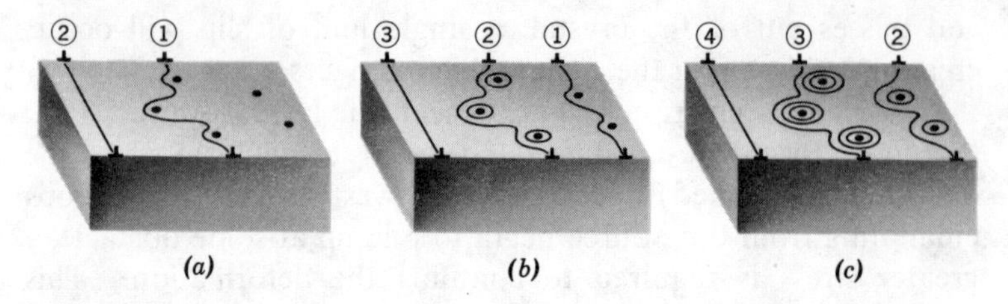


Figure 4.22 The formation of dislocation loops around precipitate particles.

## 4.9 PRECIPITATION-HARDENING

Precipitation-hardening is the strengthening of a material by the formation of a dispersion of hard, second-phase particles within the matrix, which impede the motion of dislocations through the material. A dislocation may, in passing through the matrix, arrive at a hard, high-modulus particle which does not shear as easily as the matrix. The dislocation, on being arrested in its motion at precipitate particles, starts to bulge through between them, as shown in Figure 4.22. The stress required to bulge the dislocation through is that given by Equation 4.11, $\tau = 2Gb/l$; clearly, for a given volume of second-phase precipitate, a higher stress is required, the smaller the particle, because the interparticle spacing, $l$, will be smaller. After bulging through between the particles, the dislocation line re-forms, leaving a dislocation loop around the particle (see Figure 4.22). Each additional dislocation leaves another dislocation loop around the particle, effectively increasing its size and decreasing the spacing between nearby particles, so that an ever-increasing stress is required to push successive dislocations through.

Precipitation-hardening is most effective when the particles are small and coherent with the matrix lattice. This is because a strain is created in the matrix which interacts with the strain fields of the dislocation to cause a repulsion. For a given total amount of second phase, the number of particles will be very much larger if the particles are coherent than if they are incoherent because the coherent particle can contain relatively few atoms. Also because of the relatively large number of particles, the interparticle spacing can be extremely small. The recovery processes of cross-slip and climb may operate to allow the dislocation to avoid some precipitate particles; if the dispersion is fine enough, however, too great an increase in the dislocation line length may be necessary for these processes to be energetically favorable.

A sufficient number of dislocation loops may be created around a particle after the passage of many dislocations that it will shear. This creates a crack nucleus within the matrix which often produces sudden fracture at high stress and diminishes the strain to fracture.

## 4.10 OBSERVATION OF DISLOCATIONS

The surfaces of freshly grown crystals often show spiral steps which are a result of the growth of the surface in a spiral around a screw dislocation piercing the surface. Since the lattice surrounding a dislocation is in a highly strained state, it is preferentially attacked by etchants which dissolve away surface atoms, and the resulting etch pits may be observed optically (Figure 4.23*a*). The spacing of dislocations in a low-angle grain boundary, for example, may be measured by this technique.

The tendency of foreign atoms to migrate toward dislocations makes possible the technique called *decoration,* in which the pattern of foreign atoms segregated along dislocation lines may be observed optically, in transparent crystals (Figure 4.23*b*). The electron microscope is often used for observing the pattern of dislocation entanglement resulting from plastic deformation (Figure 4.23*c*). X-ray techniques are also used, as, for example, in the measurement of the misorientation of planes across a low-angle grain boundary.

### DEFINITIONS

*Slip.* The relative sliding of parallel planes of atoms; equivalent lattice sites are again occupied by each atom after slip.

*Slip Plane.* The crystallographic planes on which slip occurs in a given crystal structure.

*Dislocation.* An imperfection in a crystal which is the boundary between the slipped and unslipped portions of the crystal.

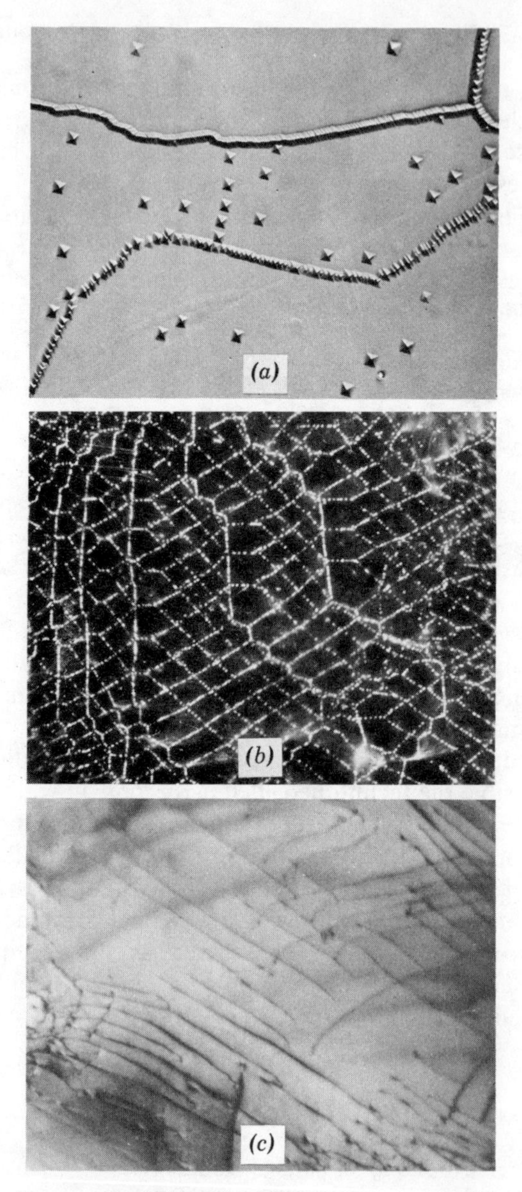


Figure 4.23 Photographs of dislocations. (*a*) Etch pits in LiF (290×). (Courtesy of W. G. Johnston, General Electric Company.) (*b*) Dislocations in sodium chloride decorated with silver (405×). (Courtesy of S. Amelinckx, S. C. K. Mol-Donk, Belgium.) (*c*) Electron transmission photograph of dislocations in niobium single crystal (11,600×). (Courtesy of C. S. Tedmon, M.I.T.)

*Burgers Vector,* **b.** A unit slip vector.
*Cross-Slip.* The change of slip plane by a screw dislocation.
*Climb.* The change to a parallel slip plane by an edge dislocation.
*Tilt Boundary.* A dislocation wall separating two regions of lattice misoriented by a relatively small angle.
*Jog.* An increase in length of dislocation line created by intersection of dislocations; often a geometric step in the line.
*Stacking Fault.* A local region across which the crystallographic packing differs from that characteristic of the crystal.
*Partial Dislocation.* A dislocation whose Burgers vector does not represent the unit of slip; the vectorial sum of Burgers vectors of partial dislocations, however, often does represent the unit of slip.
*Sessile Dislocation.* A dislocation for which the Burgers vector and the dislocation line do not both lie in an active slip system; an immobile dislocation.
*Dislocation Density.* The length of dislocation line present in a unit volume of a crystal; or, the number of dislocation lines piercing a unit area of the crystal.
*Work-Hardening.* The increase in resistance to deformation with continuing distortion.
*Precipitation-Hardening.* The increase in resistance to deformation afforded by precipitate particles which offer a resistance to dislocation motion.
*Anneal.* To soften by heating; thermal activation of processes leading to the rearrangement of existing dislocations into lower-energy configurations or the reduction of dislocation density.

## BIBLIOGRAPHY

SUPPLEMENTARY READING:

Cottrell, A. H., *The Mechanical Properties of Matter,* John Wiley and Sons, New York, 1964, Chapters 5, 9, 10, 11.

Cottrell, A. H., *Dislocations and Plastic Flow in Crystals,* Clarendon Press, Oxford, 1961, Chapters 1, 2, 3.

ADVANCED READING:

Friedel, J., *Dislocation Interactions and Internal Strains, Internal Stress and Fatigue in Metals,* Elsevier, Amsterdam, 1959.

Nabarro, F. R. N., "Mathematical Theory of Stationary Dislocations," *Advanced Physics,* **1** (1952), 269.

Seeger, A., "Mechanism of Glide and Work Hardening in Face-Centered Cubic and Hexagonal Close-Packed Metals," in *Dislocations and Mechanical Properties of Crystals,* ed. by J. Fisher et al., John Wiley and Sons, New York, 1957.

## PROBLEMS

4.1 (a) Describe a dislocation as introduced by Volterra using an isotropic elastic cylinder as a model of an edge dislocation.

(b) Use a sketch of a hollow elastic cylinder stressed in such a fashion as to produce the stress condition present in a screw dislocation.

4.2 Make a schematic representation of a dislocation pile-up at a surface or any boundary barrier.

4.3 Indicate with sketch, according to Cottrell, steps in dislocation multiplication. (*Dislocations and Plastic Flow in Crystals,* Clarendon Press, Oxford, 1953.) Show how continuous slip can occur by the motion of a single dislocation line.

4.4 Indicate with the aid of sketches the formation of growth spirals and rings on the intersection of a screw dislocation with a surface.

4.5 From what experimental measurements can the density of dislocations be deduced?

4.6 Indicate with simple sketches the following situations.

(a) The climb of a dislocation by the addition of atoms.

(b) The climb by the subtraction of atoms from an extra half plane.

(c) A screw dislocation generates an edge component in climb by formation of a helix.

4.7 (a) What do dislocation etch pits show? (b) How can dislocations in transparent ionic crystals be made visible?

4.8 For a nearly perfect crystal (near zero dislocation density) the 0.2 percent off-set yield strength may be of the order of $G/20$; at higher dislocation densities typical of well-annealed crystals ($10^6$ to $10^8$ lines/cm$^2$) the 0.2 percent offset yield strength is often very low—about $G/5000$—while with increasing dislocation density ($10^8$ to $10^{12}$ lines/cm$^2$) the 0.2 percent offset yield strength will often exhibit a value of the order of $G/100$. Explain the reason for observations of this kind.

4.9 Consider dislocations as pinned at a mean interdislocation spacing $l = \rho^{-1/2}$, where $\rho$ is the dislocation density in lines/cm$^2$ piercing a surface. Assume that the Frank-Read extrusion stress (Equation 4.11) determines the flow stress. Calculate the density of dislocations in a copper crystal strain-hardened to the stage where under an axial stress shear occurs on a particular plane, at an angle to the axial stress, on which the shear stress is 6000 psi. $G = 6 \times 10^6$ psi; $b = 10^{-8}$ in.

4.10 Calorimetric measurements of stored energy in strain-hardened copper indicate a value of the order of 3 cal/cm$^3$. Assuming that all the stored energy is due to the strain fields of dislocations and that edge and screw dislocations are present in equal numbers, calculate the dislocation density, $\rho$, in lines/cm$^2$ piercing the surface. $G = 6 \times 10^6$ psi, $b = 10^{-8}$ in.; 1 in.-lb $= 2.7 \times 10^{-2}$ cal.

4.11 Distinguish among the direction of the dislocation line, the Burgers vector and the direction of motion of the dislocation line for the edge and the screw dislocation.

4.12 The force between edge dislocations is not central, as it is between screw dislocations. From Equations 4.13 and 4.12, the radial and tangential components of force per unit length between two parallel edge dislocations having the same Burgers vector are

$$\left(\frac{F}{l}\right)_r = \frac{Gb^2}{2\pi(1-\nu)r} \qquad \left(\frac{F}{l}\right)_\theta = \frac{Gb^2 \sin 2\theta}{2\pi(1-\nu)r}$$

If the component of force, $(F/l)_x$ is taken to be in the direction of the slip plane, we can write

$$\left(\frac{F}{l}\right)_x = \left(\frac{F}{l}\right)_r \cos\theta - \left(\frac{F}{l}\right)_\theta \sin\theta$$

(a) Now show that

$$x\left(\frac{F}{l}\right)_x = \frac{Gb^2}{2\pi(1-\nu)} \cdot \left(\frac{x}{y}\right)^2 \cdot \frac{\left[\left(\frac{x}{y}\right)^2 - 1\right]}{\left[\left(\frac{x}{y}\right)^2 + 1\right]}$$

and plot $\dfrac{2\pi(1-\nu)}{Gb^2} x\left(\dfrac{F}{l}\right)_x$ versus $\dfrac{x}{y}$.

(b) Positive values of $(F/l)_x$ correspond to a repulsion force, whereas negative values correspond to an attraction force. $(F/l)_x$ is zero for $x = 0$ and for $x = y$. Explain why each of these positions is or is not a stable position.

4.13 Show that the force acting per unit length on a dislocation is $f \equiv (F/l) = \tau b$, where $l$ is the length of the dislocation and $F$ is the total force acting on it. (*Hint:* Consider a crystal containing an edge dislocation. In moving this dislocation across the crystal, the force on the dislocation moves through a distance equal to the length of the crystal while the shear force moves through a distance $|b|$.

4.14 Calculate the equilibrium radius of curvature to which a dislocation will bow out under an applied stress if it is pinned at a length $l$.

4.15 Show why a jog in an edge dislocation can glide through the lattice along with the line, whereas a jog in a screw dislocation cannot glide.

4.16 Calculate the probability of forming a jog in a dislocation in copper by thermal activation at $RT$ and at $MP$. Assume that the energy of the jog is free energy. For copper, $G = 6 \times 10^6$ psi, $b = 10^8$ in., molar volume $= 0.44$ in.$^3$, MP $= 1083°$C, probability $= e^{-Q/RT}$, 1 in.-lb $= 2.7 \times 10^{-2}$ cal.

4.17 (a) Consider a circular dislocation loop of radius $r$ spontaneously formed by thermal activation in a crystal subjected to a shear stress $\tau$. The applied stress does work $\pi\tau b r^2$ and creates a line energy $2\pi r G b^2$. Derive an expression for the minimum radius for which a stable loop is created as a function of applied stress.

(b) Calculate whether the spontaneous formation of stable loops by thermal activation is probable at reasonable values of shear stress $\tau \leq 10^{-3}G$. The affected volume is a ring of radius $2\pi r$ and transverse area $b^2$. Consider copper, for which the necessary data are given in the previous problem.

4.18 Show that the dislocation reaction in a BCC metal given by

$$\frac{a}{2}[111] + \frac{a}{2}[1\bar{1}\bar{1}] \rightarrow a[100]$$

is vectorially proper and energetically favorable.

4.19 (a) Make a schematic sketch of a dissociated screw dislocation showing the faulted region, the Burgers vectors of the two partials, and a region where the two partials have combined.

(b) Show with a sequence of sketches how the dissociated screw dislocation can pass by a hard precipitate particle in its slip plane without shearing the particle and without leaving a dislocation loop surrounding the particle.

4.20 A dislocation in an FCC lattice lies parallel to $[1\bar{1}0]$ and has the Burgers vector $a/2[10\bar{1}]$. As shown in Figure 4.17, the dislocation may dissociate:

$$\frac{a}{2}[10\bar{1}] \rightarrow \frac{a}{6}[11\bar{2}] + \frac{a}{6}[2\bar{1}\bar{1}]$$

A stacking fault of width $w$ and energy $\gamma$ erg/cm$^2$ exists between the two partial dislocations. Work is done by the force between the two partial dislocations as they move apart, but energy is required to create the stacking fault; the equilibrium value of $w$ occurs when the differential change in energy, $dE$, is zero:

$$dE = -\left(\frac{F}{l}\right) dw + \gamma \, dw = 0$$

where $F/l$ is the force per unit length between the partial dislocations given by Equation 4.12.

(a) Calculate $w/d$, the ratio of stacking fault width to atom diameter, for the following metals:

| Metal | Atom Diameter, Å | $G_{11\bar{2}}$, psi | $\gamma$, erg/cm$^2$ |
|---|---|---|---|
| Aluminum | 2.856 | $3.68 \times 10^6$ | 200 |
| Copper | 2.552 | $5.21 \times 10^6$ | 40 |
| Cu–7%Al | — | $5 \times 10^6$ | 4 |

(Take Poisson's ratio $\nu$ equal to 1/3 for all three metals.)

(b) Explain why the Cu–7%Al alloy should be expected to strain-harden at a greater rate than aluminum.

4.21 When a dislocation dissociates into a pair of partial dislocations with a stacking fault of width $w$ between them, the initial energy of the original dislocation per unit length, $E_1 \simeq Gb^2$, is lowered by the amount

$$E_2 = E' + \int_d^w \left(\frac{F}{l}\right) dw,$$

the work done by the repulsion force per unit length, $F/l$ (Equation 4.12) between the partial dislocations, and the energy is increased by the amount $E_3 = \gamma w$, where $\gamma$ is the stacking fault energy, erg/cm$^2$. (Notice that the lower limit of the integral is taken as the atom diameter, $d$, to avoid an infinite value of the integral—see pages 67–68; $E'$ is the work done in separating to the distance $w = d$ and is finite.)

(a) For the dislocation dissociation given in Problem 4.14, sketch curves $E_1 - E_2$, $E_1 + E_3$, and the curve $E_4 = E_1 - E_2 + E_3$ on coordinate axes having stacking fault width, $w$, as abscissa, and energy per unit length as ordinate; take the zero of energy to be the perfect lattice and $E_1$ is then the intercept of $E_1 - E_2$, $E_1 + E_3$ and $E_1 - E_2 + E_3$ on the ordinate. Show the quantity $E'$ and the equilibrium stacking fault width, $w_0$, on the graph. What geometric relationship exists between the curves $(E_1 + E_3)$ and $(E_1 - E_2)$ at the equilibrium spacing?

(b) The energy per unit length of the initial dislocation is of the order of $E_1 \simeq Gb_1^2$ (Equation 4.6) and at the equilibrium spacing, $w_0$, two partial dislocations exist, each having energy $E_p \simeq Gb_p^2$. Then $E_4 = E_1 - E_2 + E_3 = 2E_p$ at equilibrium. Estimate the ratio $E'/E_1$ for brass for which the ratio of stacking fault width to atom diameter is $w/d = 22$.

4.22 A severely strain-hardened sample of polycrystalline copper has a dislocation density of $10^{12}$ cm/cm$^3$; when thoroughly annealed, the same sample has a dislocation density of $10^7$ cm/cm$^3$.

(a) Calculate, in cal/cm$^3$, the energy stored by strain-hardening of an annealed specimen of this copper to $\rho = 10^{12}$ cm/cm$^3$. (Mean value of $G = 6 \times 10^6$ psi; $b \simeq 10^{-8}$ in.; take $\nu = \frac{1}{3}$; assume that the dislocations have 50 percent edge and 50 percent screw character, and that, because of the presence of stacking faults, the energy of the dislocations is only 40 percent of what it would be if they were undissociated.)

(b) To what height, in feet, would 1 cm$^3$ of copper have to be lifted to increase its potential energy by the same amount as its internal energy is increased by cold-working to $\rho = 10^{12}$ cm/cm$^3$? (Density of copper = 8.96 g/cm$^3$.)

4.23 Read pp. 66–70 in Cottrell "Dislocations and Plastic Flow in Crystals."

(a) State briefly what is meant by dislocation damping.

(b) Discuss how the elastic modulus of an isotropic single crystal containing a single edge dislocation should vary as a function of the frequency of the applied stress.

CHAPTER FIVE

# *Microplasticity of Crystals*

Deformation in crystalline materials occurs principally by slip along specific planes and directions. Twinning may contribute to the deformation when slip is constrained. At high temperatures and low rates of deformation, polycrystalline materials may also deform by grain boundary sliding or diffusional creep.

## 5.1 INTRODUCTION

The resistance of a crystalline body to elastic or plastic deformation is dependent on the bonding forces between the atoms. However, the discrete processes which produce plastic strain involve the point, line, and surface *imperfections* discussed in Volume I, Chapter 4.

The most important mechanism of crystal deformation is *slip,* the gliding of planes of atoms over one another, which occurs by the motion of dislocations over specific crystallographic planes and in definite crystallographic directions. The process of mechanical *twinning,* although generally of less importance, can contribute significantly to plastic strain if slip is restricted. Twinning is the conversion of a portion of the crystal into a mirror image of the parent lattice by homogeneous shearing of successive lattice planes by the amount of the *twinning vector.* The shear strain is equal to the magnitude of the twinning vector divided by the interplanar spacing.

In polycrystalline materials, two additional deformation processes may operate at low stresses and temperatures near the melting point, *grain boundary sliding* and *diffusional creep.* During grain boundary sliding, whole grains are displaced relative to one

Figure 5.1 Slip lines in copper-2 percent aluminum (850×). (Courtesy of G. Miller, M.I.T.)

another to the extent allowed by slip accommodation at the corners were grains intersect. In diffusional creep, the individual grains are elongated in the direction of maximum tensile stress by a flux of self-diffusing atoms. The contributions of each of these processes to plastic deformation, and the role of imperfections in the strain-hardening and recovery of crystals during deformation, are emphasized in the present chapter.

## 5.2 SLIP PLANES AND SLIP DIRECTIONS

One or more sets of fine parallel lines may be observed to traverse the surface of a crystal examined after plastic straining (see Figure 5.1). At higher magnification, these *slip lines* are seen to result from the relative displacement of crystal planes by distances of 20 to 500 atom diameters. At low temperatures, continued plastic deformation creates new slip lines, and not continued slip along existing lines, indicating that the plane is made more resistant to shear as the result of the slip process. At higher temperature, however, slip lines cluster into coarser bands, and very little slip occurs between the bands. Individual lines of the band glide repeatedly, and thus develop large local steps in the crystal surface called *slip bands* (Figure 5.2).

Slip lines are straight in FCC and HCP materials. In BCC materials, however, the lines are extremely wavy and are apparently not produced by glide on a single plane. The active slip planes and the directions in these planes along which displacement occurs have been determined by two methods—measurement of the angle of slip lines on crystal surfaces of known orientation, and correlation of the total displacement on the operating slip directions and planes with the total crystal strain. The observed slip planes and directions in FCC, BCC, and HCP metals and FCC ionic crystals are listed in Table 5.1. The first-listed plane in each case is the one most commonly observed. The combination of a slip direction and the plane containing it is defined as a *slip system.*

In each case, the slip direction is the most closely packed direction of the lattice. Since slip results from dislocation motion, those dislocations which require the lowest energy will move most

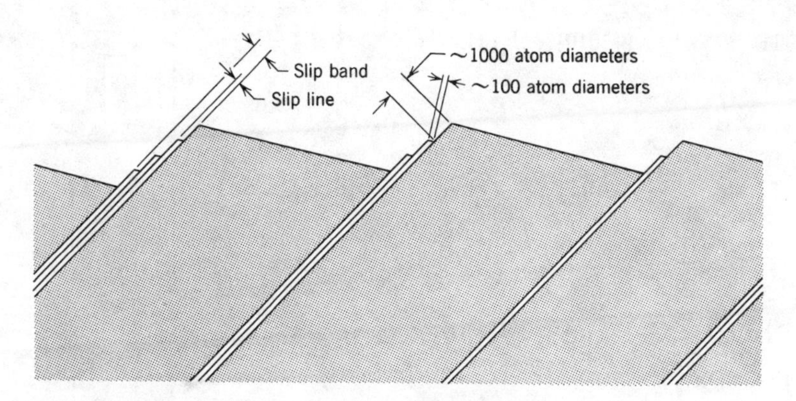


Figure 5.2 Profiles of slip bands.

*Table 5.1 Observed Slip Systems in Crystals*

| STRUCTURE | SLIP PLANE | SLIP DIRECTION | NUMBER OF SLIP SYSTEMS |
|---|---|---|---|
| FCC<br>Cu, Al, Ni,<br>Pb, Au, Ag,<br>γFe, ... | {111} | $\langle 1\bar{1}0 \rangle$ | $4 \times 3 = 12$ 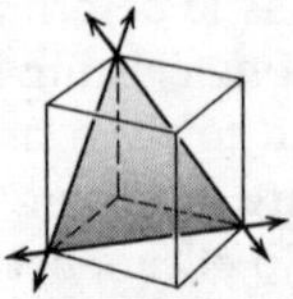 |
| BCC<br>αFe, W, Mo,<br>β Brass | {110} | $\langle \bar{1}11 \rangle$ | $6 \times 2 = 12$ 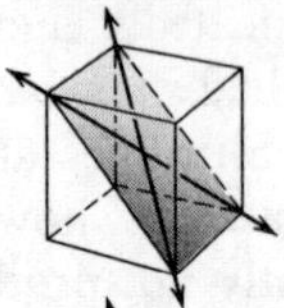 |
| αFe, Mo,<br>W, Na | {211} | $\langle \bar{1}11 \rangle$ | $12 \times 1 = 12$ 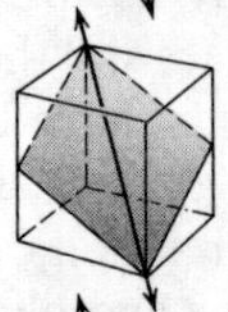 |
| αFe, K | {321} | $\langle \bar{1}11 \rangle$ | $24 \times 1 = 24$ 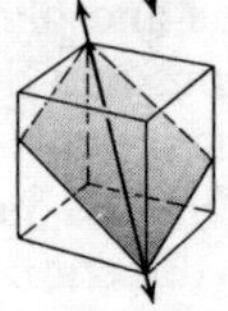 |
| HCP<br>Cd, Zn, Mg,<br>Ti, Be, ... | (0001) | $\langle 11\bar{2}0 \rangle$ | $1 \times 3 = 3$ 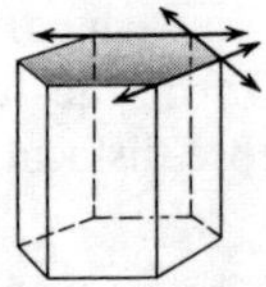 |
| Ti | $\{10\bar{1}0\}$ | $\langle 11\bar{2}0 \rangle$ | $3 \times 1 = 3$ 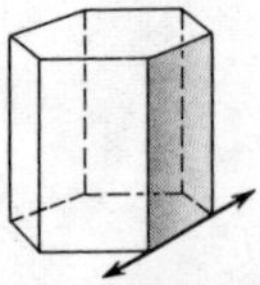 |
| Ti, Mg | $\{10\bar{1}1\}$ | $\langle 11\bar{2}0 \rangle$ | $6 \times 1 = 6$ 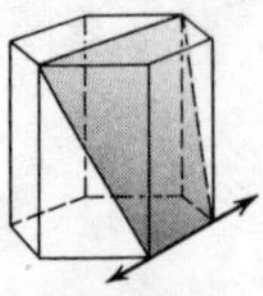 |
| NaCl, AgCl | {110} | $\langle 1\bar{1}0 \rangle$ | $6 \times 1 = 6$ 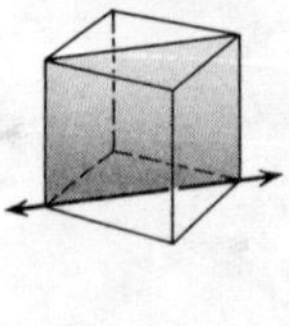 |

readily. As the energy of a dislocation is proportional to the square of the Burgers vector, dislocations of the shortest Burgers vector, that is, those along the close packed direction, will require the least stress.

The slip planes are usually the most densely packed planes, which are also the most widely separated. A higher shear stress is required to produce slip on planes of lower packing density. However, if slip on the closely packed planes is constrained by such conditions as high local stress due to strain-hardening, the restriction of slip by grain boundaries, etc., these latter planes may become active.

In FCC metals, only one type of slip system has been observed, $\{111\}\langle 1\bar{1}0\rangle$. The total strain in the crystal is accommodated by slip on one or more of the four $\{111\}$ planes. As each plane contains three $\langle 1\bar{1}0\rangle$ slip directions, there are twelve possible slip systems. The most important slip system in BCC metals is the $\{110\}\langle \bar{1}11\rangle$. However, fine and usually slightly wavy slip traces are commonly observed on the $\{211\}$ and $\{321\}$ planes. Comparison of effects of crystalline slip with the total strain suggests either that slip is occurring simultaneously on several $\langle 110\rangle\langle \bar{1}11\rangle$ systems or that the other slip systems are active.

The number of slip systems in the HCP metals is smaller. Three systems lie in the basal and most closely packed plane, $(0001)\langle 11\bar{2}0\rangle$; these are the normally active systems in Zn and Cd. Other systems are operative at elevated temperatures. Gilman has observed prismatic flow in these crystals at high temperatures. The ratio of the lattice parameters plays an important role in determining whether other slip systems are possible. For ideal packing of spheres in HCP configuration the $c/a$ ratio is 1.632. However, this ideal ratio does not occur in any HCP metal. In Zn and Cd, $c/a$ is considerably higher than the ideal (1.856 and 1.886). In the other HCP metals, $c/a$ is lower than the ideal: Mg (1.624), Ti (1.587), Be (1.568), Zr (1.59), Hg (1.586), Re (1.617), and Co (1.624). When the $c/a$ ratio is low, the (0001) plane loses the distinction of being the plane of highest atomic density. Thus Ti can show additional slip planes, and in Mg the $\{10\bar{1}1\}$ plane is reported to contribute to deformation above 210°C. In each case, the slip direction remains $\langle 11\bar{2}0\rangle$.

## 5.3 RESOLVED SHEAR STRESS

Crystalline slip results from the action of a shear stress on the slip plane. Within the range of stresses in engineering application, the component of stress normal to the slip plane does not influence slip. Thus the slip process must be considered in terms of the shear stress resolved on the slip plane in the slip direction. Consider a single crystal of cross-sectional area $A$ under a tensile force $F$ (Figure 5.3). Let $\phi$ be the angle between the slip plane normal and the tensile axis, and $\lambda$ the angle between the slip direction and the tensile axis. The component of the applied force, acting in the slip direction is $F \cos \lambda$, and the area of the slip plane is $A/\cos \phi$. The shear stress resolved in the slip direction is then

$$\tau = \frac{F \cos \lambda}{A/\cos \phi} = \sigma \cos \phi \cos \lambda \tag{5.1}$$

where $\sigma$ is the applied tensile stress $F/A$.

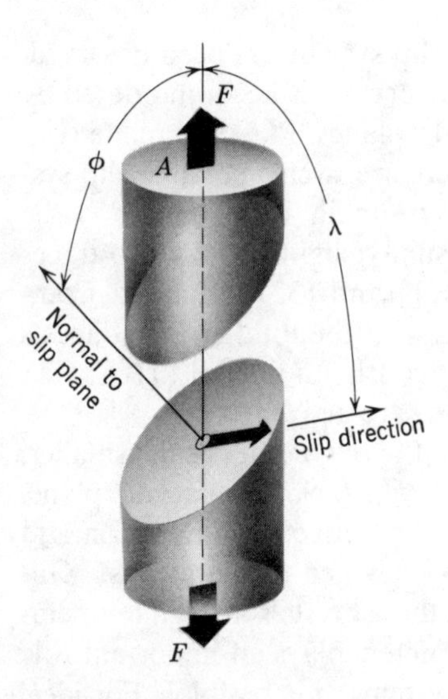


Figure 5.3 Geometry used for calculating resolved shear stress.

The stress required to initiate slip in a pure and perfect single crystal, the *critical resolved shear stress,* is a constant for a material at a given temperature. This rule, known as Schmid's Law, has been experimentally proven for a large number of metal single crystals. Curves of the critical stress required to cause yielding, as a function of $\cos \phi \cos \lambda$, are shown in Figure 5.4 for two values of critical resolved shear stress. If many slip systems of the same type are possible in a crystal, the active plane is the one on which the critical resolved shear stress is reached first as the specimen is subjected to increasing applied stress.

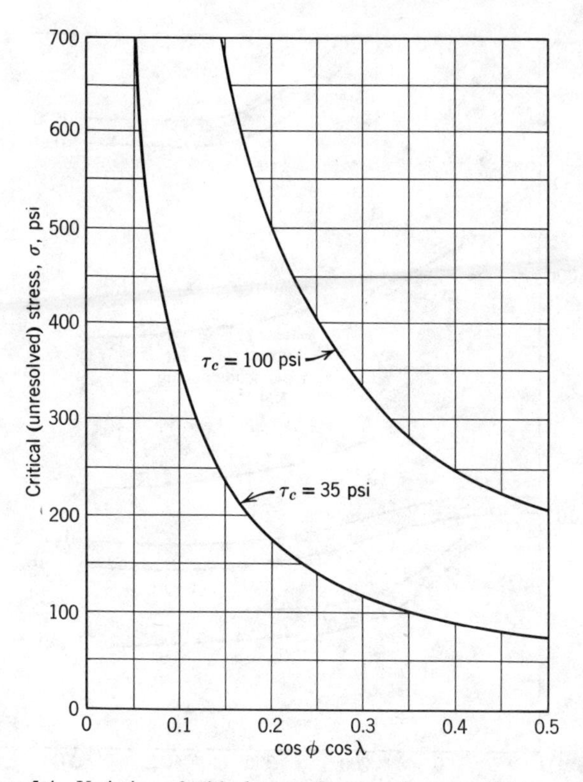


Figure 5.4 Variation of critical unresolved tensile stress with cos $\phi$ cos $\lambda$.

Room-temperature values of critical resolved shear stress vary over a wide range. The critical resolved shear stress of such FCC metals as aluminum and copper and such HCP metals as zinc and magnesium are of the order of 100 psi, whereas in BCC metals such as iron and tungsten they are typically several ten thousand psi. Certain ionic crystals of the NaCl structure exhibit intermediate values. Figure 5.5 illustrates the variation of critical resolved shear stress with temperature for several materials; note that the difference in its magnitude increases markedly at lower temperature. The exact value of critical resolved shear stress is strongly dependent on the purity of the material and the perfection of the lattice; as a consequence, different values are often

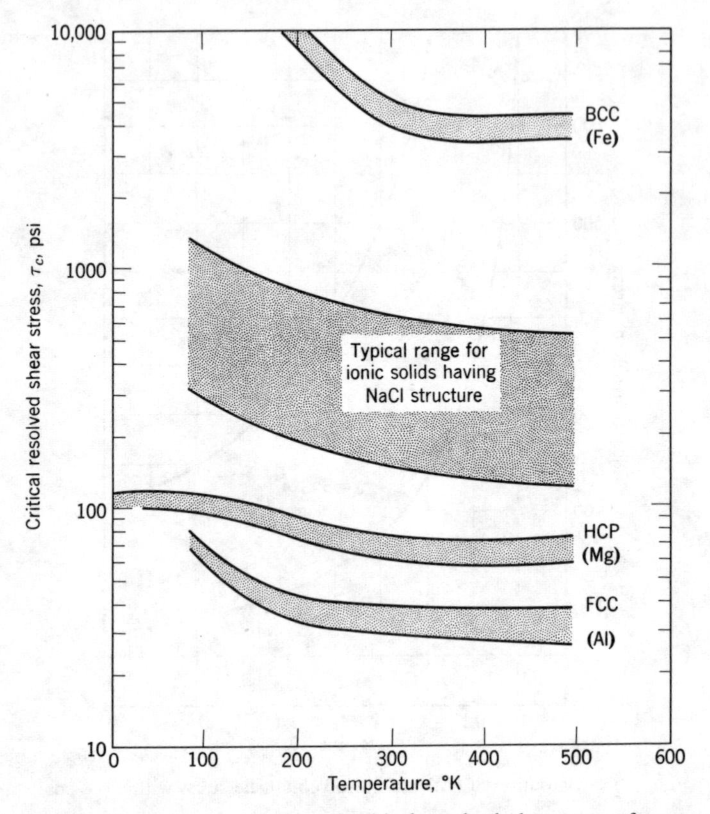


Figure 5.5 Temperature variation of critical resolved shear stress for several classes of crystalline materials.

found for two materials which are nominally the same. Figure 5.5 shows the approximate range in values for the several kinds of materials.

## 5.4 STRAIN-HARDENING AND RECOVERY OF SINGLE CRYSTALS

When a crystalline material is plastically deformed, it becomes stronger; it work- or strain-hardens, and an increased stress is required for further strain. The slope of the stress-strain curve is used to define the rate of strain-hardening. Shear stress-shear strain curves (shear stress and strain resolved on the active slip system) for Mg, Cu, and Fe single crystals are shown in Figure 5.6 for orientations such that the resolved shear stress will be greatest on a single slip system over the major portion of the deformation. Under this condition the majority of slip will occur on only one slip system in HCP and FCC metals, even though the applied stress, by strain-hardening, becomes sufficiently large that the critical resolved shear stress is exceeded on a second, less favorably oriented system. A phenomenon known as *latent hardening* results in raising of the resolved flow stress for shear on all slip systems of a type when any one system is strengthened by strain-hardening. If the crystal is oriented so that the resolved shear stresses on two or more slip systems are equal, then these systems will slip simultaneously. The several slip system traces seen in Figure 5.1 arise in part from this *multiple slip* process.

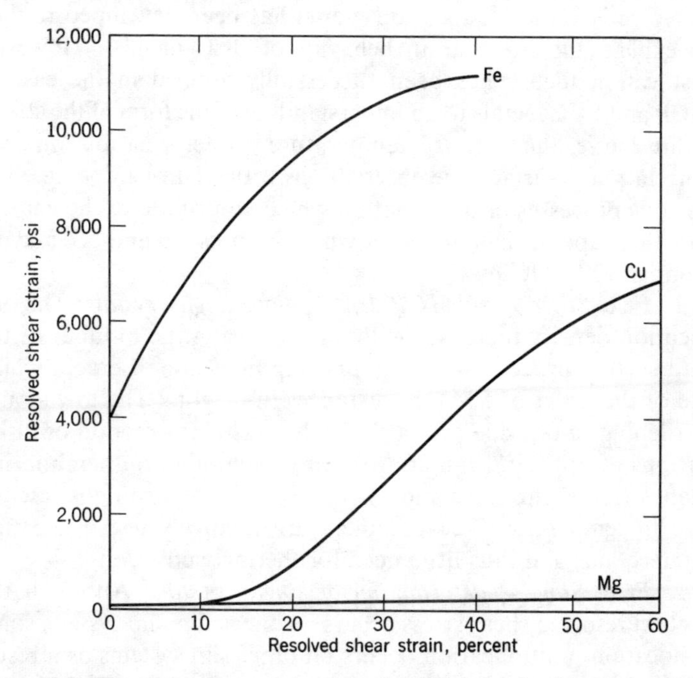


Figure 5.6 Shear stress-shear strain curves for single crystals of iron, copper, and magnesium.

The BCC metals, however, do not behave in the same fashion as the HCP and FCC metals. From the start of deformation, the BCC crystal will slip on two (or more) planes which share the direction of maximum stress rather than on a single plane.

As illustrated in Figure 5.6, the stress-strain curve for HCP magnesium shows an essentially constant low hardening rate to fracture, whereas the FCC copper crystal displays three stages of hardening: Stage 1, referred to, pictorially, as *easy glide,* shows the same low strain-hardening rate as the HCP crystal; Stage 2, over which the slope of the stress-strain curve is high and constant, is *linear hardening;* and Stage 3, in which the rate of strain-hardening decreases, is called *dynamic recovery.* The extent of strain in Stages 1 and 2 is sensitive to crystal orientation and decreases with increasing temperature. The BCC iron crystal, on the other hand, exhibits a decreasing strain-hardening rate over the entire range of deformation.

No satisfactory dislocation model has been developed to date to explain the stress-strain behavior of BCC metals. However, dislocation theory has been successfully applied in the case of HCP and FCC metals to an understanding of the form of the stress-strain curve, the effect of temperature on deformation, and the correlation of surface line observations with strain-hardening. The various processes of deformation which contribute to the characteristic shape of the curves shown in Figures 5.6 and 5.7 may be summarized as follows:

1. *FCC Stage 1 and HCP deformation, single crystals:* The dislocation density increases without increasing the number of the intersection processes. Long, fine slip lines are observed, which are of the order of 50 atom distances in height. The low strain-hardening rate is due primarily to the elastic interaction of dislocations of opposite sign approaching each other on neighboring planes (see Figure 5.8*a* and *b*). The rate of hardening is insensitive to temperature because there are relatively few dislocation intersections and thus little need for thermal activation.

2. *FCC Stage 2 deformation of single crystals:* Although the highest resolved shear stress occurs on one active slip system, some dislocation multiplication occurs on other slip systems as a result of local stresses at dislocations which become stuck on the active slip system. Dislocations on the active system interacting with

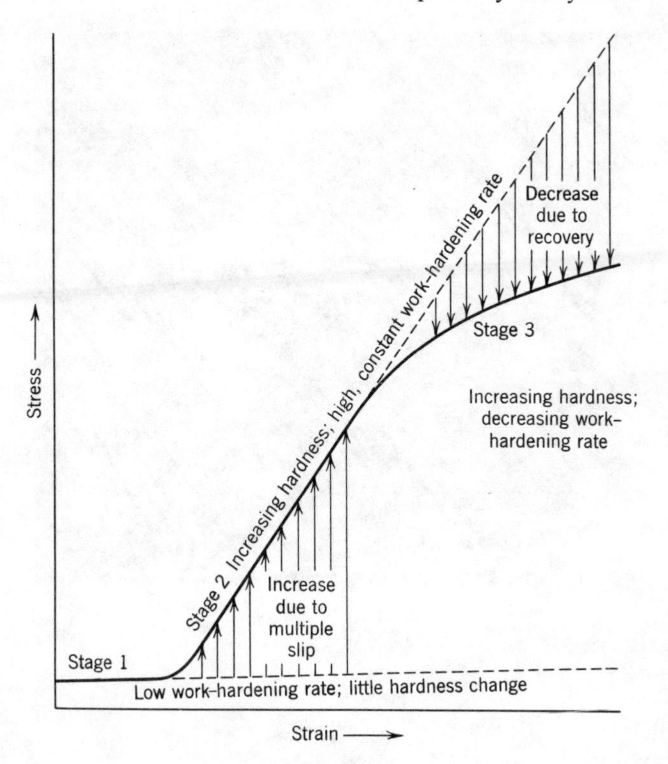


Figure 5.7 Three steps in the generic work-hardening curve for a single crystal.

those on other systems may form temporary barriers to easy motion, by intersection, or permanent barriers by the formation of *sessile* or nonmoving dislocations in the slip plane. These barriers increase in number with increasing strain and, as a result, surface slip lines become shorter, smaller in separation, and less regularly spaced. With increasing strain, a greater number of dislocations remain stuck within the crystal, raising the local internal stress, and reducing the plastic strain for an incremental increase in applied stress.

3. *Screw dislocation motion, cross-slip and FCC Stage 3 deformation, single crystals:* Screw dislocations lying entirely in the slip plane can move as easily as edge dislocations. However, when a moving screw dislocation intersects another screw dislocation on

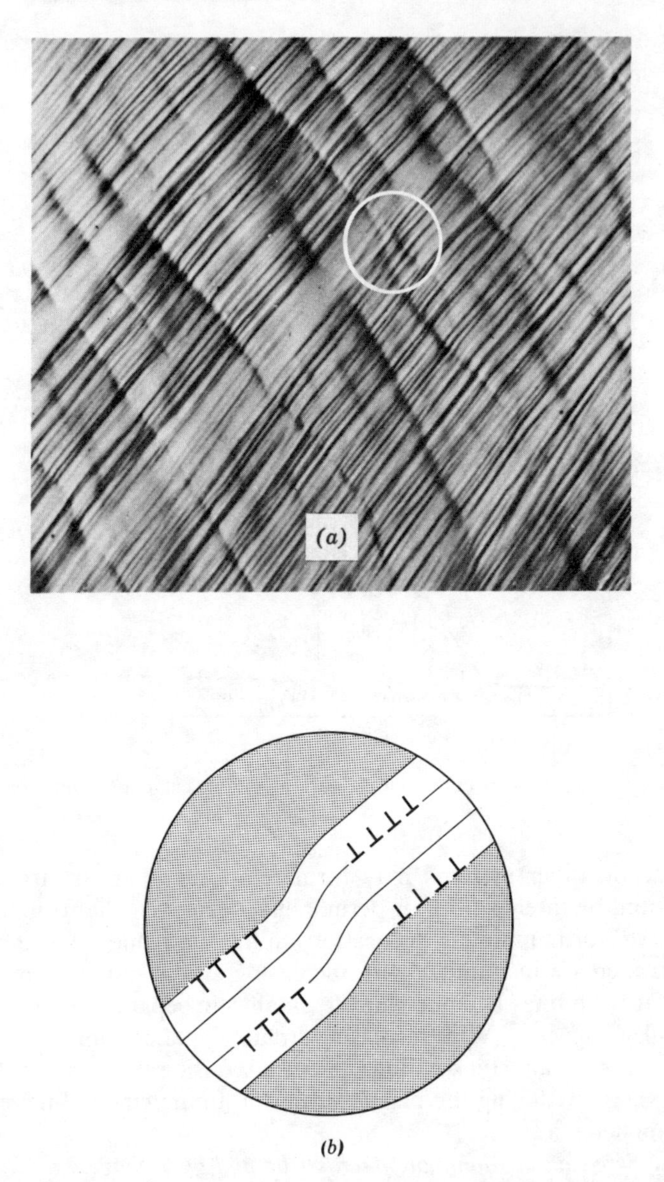


Figure 5.8 (*a*) Stage 1 work-hardening in a single crystal of copper (250×) (Courtesy of S. Mader.) (*b*) Schematic diagram of area circled in white in (*a*).

a nonactive slip system, a short edge dislocation jog is produced at the intersection. As the Burgers vector of the jog does not lie in its slip plane, it can move only by climb (i.e., by the production of interstitials or vacancies)—see Figure 4.17. Thus at lower temperatures, where the *equilibrium vacancy concentration* is low, such jogs inhibit the motion of screw dislocations as the point imperfections must be produced *athermally*. Therefore dislocations tend to form in long, narrow loops with short, rapidly moving edge dislocations at the ends and long, slower-moving screw dislocations at the edges. The motion of jogged screw dislocations is facilitated by higher stresses and higher temperatures.

Because the Burgers vector of a screw dislocation is parallel to its length, a pure screw dislocation may move from its own to a second slip system sharing the same slip direction. This process is known as *cross-slip* (Figure 4.10*a*). By cross-slip, a dislocation may bypass an obstacle in its slip plane, making further deformation easier.

By the time one gets to Stage 3 the stress buildup is so high that cross-slip is activated due to the recombination of partial dislocations. In FCC single crystals, the strain-hardening rate then continually decreases as a result of stress-induced cross-slip. Evidence of cross-slip may be seen on polished surfaces as shown in Figure 5.9. The stress at which Stage 3 is initiated decreases with increasing temperature for two reasons: (1) jogged screw dislocations move with greater ease at higher temperatures, and (2) cross-slip, which is a thermally activated process, can occur with greater frequency.

The entire stress-strain curve for a BCC crystal resembles Stage 3 deformation in FCC crystals, with a decreasing strain-hardening rate. Cross-slip is observed from the start of deformation in BCC crystals. It can occur, in large part, because the glide dislocations are whole dislocations and no recombination of partials is required for cross-slip.

4. *Recovery:* This generic name refers to processes resulting in a reduction of flow stress. Any process leading to reduction in number of dislocations by annihilation, or to the rearrangement of dislocations into lower-energy configurations, such as subgrain boundaries, falls into this category. The two major processes by

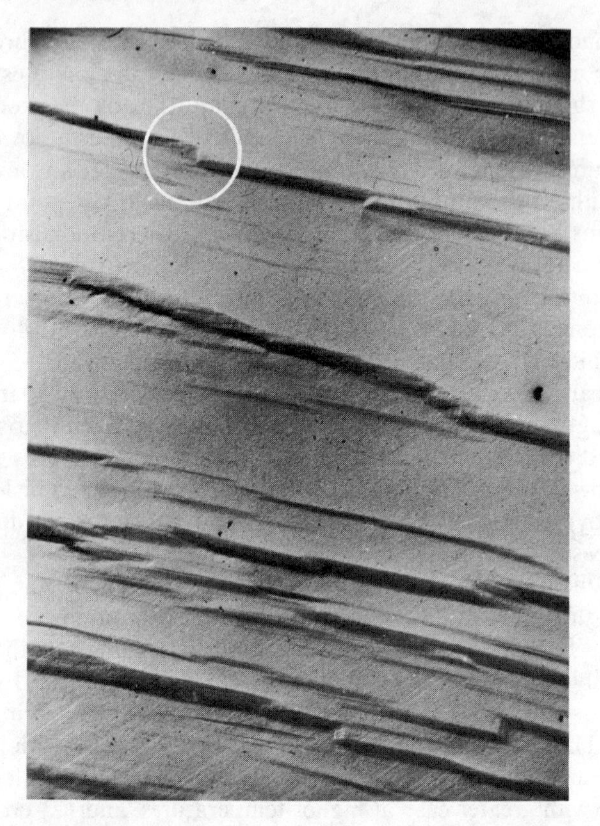

Figure 5.9 Cross-slip in a single crystal of copper (5000×). (Courtesy of S. Mader.)

which recovery occurs are *cross-slip* and *dislocation climb,* which have been discussed in Chapter 4. Cross-slip is essentially a dynamic process, that is, the largest part of the energy required must be supplied by an external stress. By cross-slipping out of its current slip plane, a dislocation can circumvent obstacles. Furthermore, cross-slipping screw dislocations often meet screw dislocations of opposite sign on adjacent planes and the two annihilate to leave a local region of perfect lattice. This reduces the dislocation content and, therefore, the hardness of the crystal.

By contrast, dislocation climb is primarily a thermally activated process which requires the migration of vacancies. The rate at which recovery by dislocation climb occurs is proportional to three thermal frequency factors: (1) the rate of formation of vacancies, (2) the equilibrium rate of formation of jogs, and (3) the rate of movement of vacancies to the jog.

In summary, the low hardening rate in Stage 1 deformation is associated with the fact that there is an absence of dislocation intersections. The high strain-hardening rate in Stage 2 is the result of the intersection and trapping of dislocations within the crystal, and the decreasing strain-hardening rate in Stage 3 is the result of dynamic recovery processes occurring simultaneously with strain-hardening.

## 5.5 TWINNING

A crystal is said to be twinned when one portion of its lattice is a mirror image of the other (see Figure 5.10). The crystallographic plane of reflection is known as the twin plane. Twins may be formed during the growth of the crystal or may be produced by mechanical twinning, which occurs by a homogeneous shear of successive planes of atoms by the amount of the twinning vector, parallel to the twin plane. Twin planes, vectors, and the

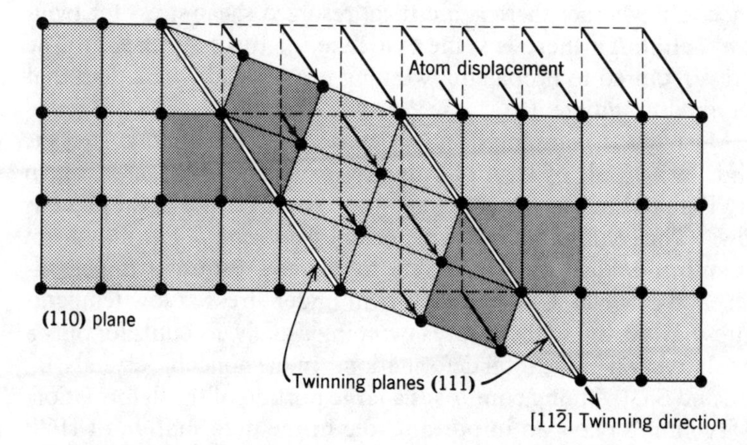


Figure 5.10 Schematic diagram of the twinning process in an FCC lattice.

*Table 5.2 Twinning Planes and Directions*

| CRYSTAL STRUCTURE | TWIN PLANE AND DIRECTION | | TWINNING SHEAR | |
|---|---|---|---|---|
| FCC | (111) | [112] | | 0.707 |
| BCC | (112) | [111] | | 0.707 |
| HCP | $(10\bar{1}2)$ | $[10\bar{1}\bar{1}]$ | Cd | 0.171 |
| | | | Zn | 0.139 |
| | | | Mg | 0.129 |
| | | | Ti | 0.189 |
| | | | Be | 0.199 |

shear produced are given in Table 5.2 for FCC, BCC, and HCP crystals.

Mechanical twinning differs from slip in the following ways: (1) the twinned portion of a grain is the mirror image of the original lattice, whereas the slipped portion of a grain has the same orientation as the original grain; (2) slip consists of a shear displacement of an entire block of the crystal, whereas twinning is a uniform shear strain; (3) the direction of slip may be either positive or negative, while the direction of shear in twinning is limited to that which produces the twin image.

The stress required to produce twinning tends to be higher and less sensitive to temperature than that necessary for slip. It is still uncertain whether there is a critical resolved shear stress for twinning, although there is some evidence for this hypothesis. The stress required to propagate twinning is appreciably less than that required to initiate it.

Mechanical twinning usually occurs when the applied stress is high as a result of strain-hardening or low temperatures, or, in HCP metals, when the resolved shear stress on the basal plane is low. Thus thin lamellar twins called *Neumann bands* may form in an iron loaded rapidly at very low temperatures. Copper, silver, and other FCC metals also twin under stress at low temperatures. However, in these cases twinning usually accounts for only a small fraction of the total deformation. In the noncubic crystals, Bi, Sb, and Sn, twinning comprises a large portion of the deformation.

Twinning plays an important role in the deformation of HCP metals, although the amount of shear it produces is small. Since slip can occur only on the basal plane in many of these metals, twinning can both contribute to the bulk deformation itself and, more important, reorient the lattice more favorably for basal slip. The HCP metals, the most common twinning plane and direction is $(10\bar{1}2)$ and $[10\bar{1}\bar{1}]$. Whether twinning will result in compression or extension along the $c$ axis is determined by the $c/a$ ratio. Zn and Cd, with a $c/a$ ratio greater than $\sqrt{3}$, will twin on $(10\bar{1}2)$ $[\bar{1}011]$ when compressed along the $c$ axis. When $c/a = \sqrt{3}$, the twinning shear is zero, and for lower $c/a$ ratios, the sense of the shear necessary for twinning is reversed. Thus Mg and Be twin under a tensile stress along the $c$ axis, as is shown in Figure 5.11.

Twinning is believed to occur by a dislocation mechanism, although twinning dislocations have not been identified experimentally. Such a process would differ from that for slip in two ways: (1) the Burgers vector of a twinning dislocation does not produce a unit lattice translation and this would not bring the lattice back in register; (2) each plane above the twin plane is displaced by a single twin vector. In the mechanisms proposed for twinning in FCC and BCC crystals, the twinning dislocation is one part of a dissociated slip dislocation which can spiral upwards over successive planes when pinned at a screw dislocation normal to the slip plane. The source of this twinning dislocation is most easily visualized in an FCC crystal, for the slip and twin planes are both $\{111\}$ planes. A slip dislocation of type $a/2\ [10\bar{1}]$ dissociates into two partial dislocations (Figure 5.12): an $a/6\ [2\bar{1}\bar{1}]$ partial which remains on the slip plane, and an $a/6\ [11\bar{2}]$ dislocation which can produce twinning.

The shear deformation associated with twinning causes large local stresses when the twin ends within the crystal. The formation of a twin is often accompanied by slip and bending in surrounding regions of the crystal and in polycrystalline metals may cause twinning in neighboring grains.

## 5.6 GRAIN-BOUNDARY SLIDING AND DIFFUSIONAL CREEP

At high temperatures and low strain rates, sliding occurs along the grain boundaries in polycrystalline materials. As grain boundaries at 45° to the tensile axis slide the most, the displacement is

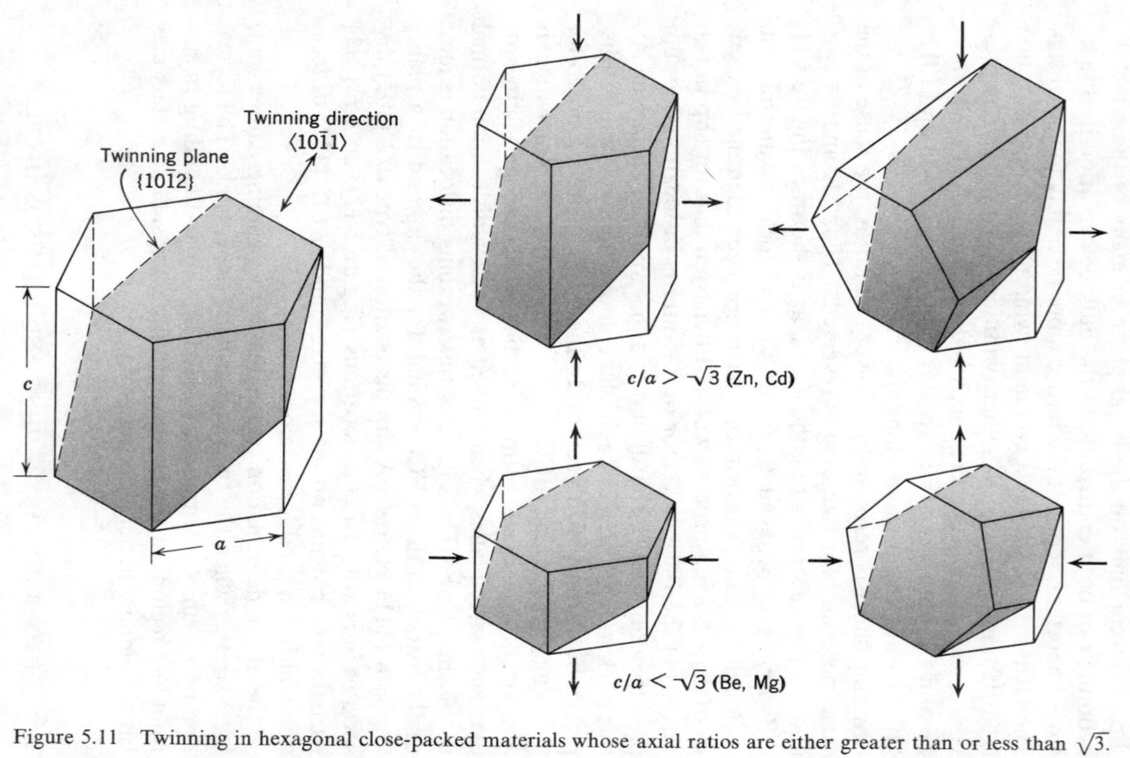


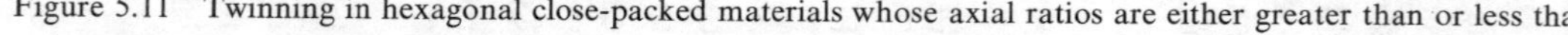

Figure 5.11 Twinning in hexagonal close-packed materials whose axial ratios are either greater than or less than $\sqrt{3}$.

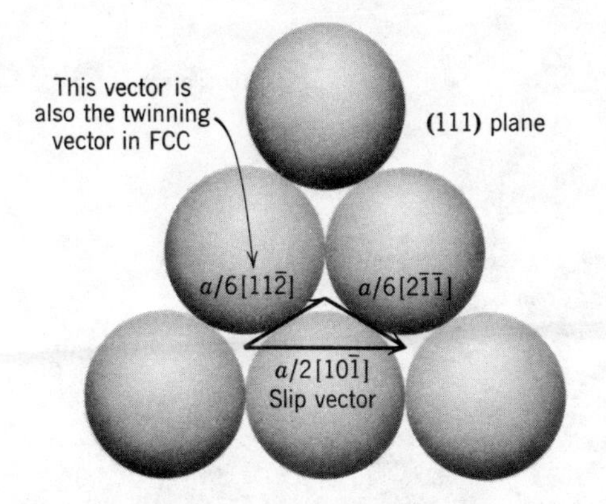


Figure 5.12 FCC twinning vector is identical with the partial slip vector.

evidently caused by the stress and contributes to the total strain (Figure 5.13). It has been found in creep tests on pure metals at low stresses that as much as 30 percent of the total strain may arise from grain boundary shear; this fraction decreases with increasing stress.

The ratio of strain due to grain boundary sliding to the total strain is constant over the course of deformation. This constancy arises because the sliding is controlled by slip processes (Figure 5.13) and accommodation of sliding is particularly difficult at grain intersections. Thus grain-boundary sliding is limited by the rate at which the crystal can adjust to this inhomogeneous deformation by normal slip processes.

At temperatures close to the melting point, where the equilibrium concentration of vacancies is high and self-diffusion is rapid, polycrystalline materials may deform by a diffusional creep mechanism rather than by slip. As shown in Figure 5.14, a grain can be elongated by the diffusion of atoms to boundaries normal to the tensile axis and vacancies to boundaries parallel to it. Thus the diffusion path is of the order of the grain size. Diffusional creep is distinguishable from deformation by slip by the fact that the rate of creep, not the amount of deformation produced, is directly proportional to the applied stress, as in Newtonian viscosity. Hence the process is sometimes referred to as the viscous flow of a crystalline material.

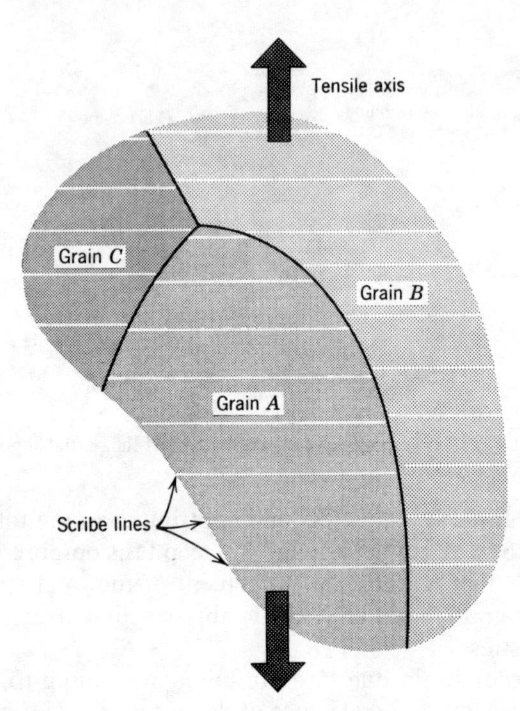


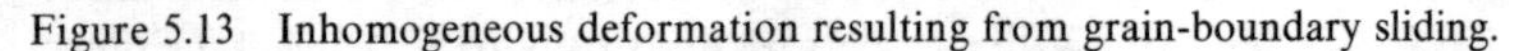
Figure 5.13 Inhomogeneous deformation resulting from grain-boundary sliding.

## DEFINITIONS

*Slip.* The parallel movement of two adjacent crystal planes relative to one another.

*Slip System.* The combination of a plane and a direction lying in the plane along which slip occurs.

*Critical Resolved Shear Stress.* The resolved stress on an active slip system, at which slip is initiated.

*Schmid's Law.* The critical resolved shear stress is a material constant at constant temperature and constant strain rate.

*Latent Hardening.* The increased resistance to slip which is developed on all *inactive* slip systems during the deformation of a crystal or grain by slip on one or more *active* slip systems.

*Easy Glide (Stage 1).* The linear, low work-hardening rate portion of a single crystal stress-strain curve characteristic of HCP crystals or of the initial deformation of FCC crystals oriented for slip on one slip system.

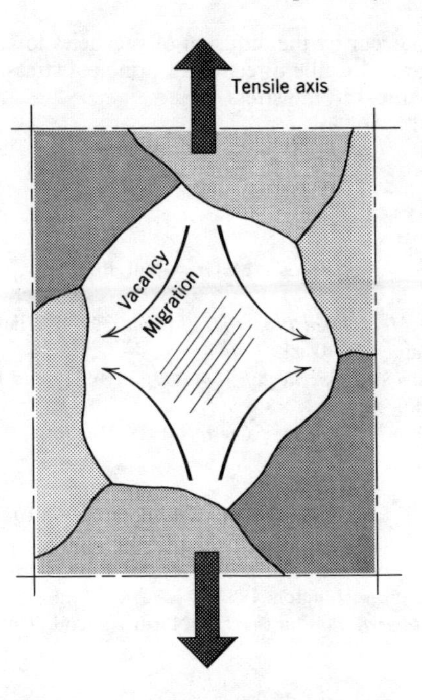


Figure 5.14 Vacancy migration to grain boundaries parallel to an applied tensile stress generates a time-dependent component to strain at constant stress.

*Linear Hardening (Stage 2).* The linear, high work-hardening portion of a FCC single crystal stress-strain curve which follows Easy Glide.

*Parabolic Hardening (Stage 3).* Deformation characteristic of FCC or BCC single crystals at high stress levels, in which the log stress is proportional to log strain.

*Recovery.* A process causing a reduction of either the flow stress or the rate of work-hardening of a deformed single crystal or polycrystal.

*Twinning.* A homogeneous shear which reorients the deformed lattice into a mirror image of the parent lattice across the plane of twinning.

*Plane of Reflection.* A symmetry plane across which the lattice is reproduced completely and identically as if it were reflected from the opposite side (e.g., a {100} plane in a cubic crystal).

*Grain-Boundary Sliding.* Prominent shear along or very near and parallel to the boundary between two grains.

*Twinning Vector.* The lattice vector representing the amount and direction of lattice displacement which occurs during mechanical twinning.

*Diffusional Creep.* A low strain rate, high temperature deformation process thought to occur by the diffusion of vacancies to grain boundaries or surfaces parallel to the direction of principal stressing. The strain rate is proportional to the stress.

## BIBLIOGRAPHY

SUPPLEMENTARY READING:

Barrett, C. S., *Structure of Metals,* McGraw-Hill Book Co., New York, 1954, Chapter 15.

Cottrell, A. H., *The Mechanical Properties of Metals,* John Wiley and Sons, New York, 1964, Chapters 9, 10, 11.

McLean, D., *Mechanical Properties of Metals,* John Wiley and Sons, New York, 1962, Chapter 4.

Schmid, E. and W. Boas, *Plasticity of Crystals,* F. A. Hughes and Co., 1950.

ADVANCED READING:

Fisher, J. C. (ed.), *Dislocations and Mechanical Properties in Crystals,* John Wiley and Sons, New York, 1951.

Hill, E. O., *Twinning and Diffusionless Transformations in Metals,* Butterworth Scientific Publications, London, 1954.

van Bueren, H. G., *Imperfections in Crystals,* North-Holland, Amsterdam, 1960.

## PROBLEMS

5.1 (a) In what type of metallic crystal lattice is ductility most readily found?

(b) On what crystallographic planes does slip take place in (1) BCC metals, in (2) FCC metals, in (3) HCP metals?

5.2 Draw a curve showing recovery, recrystallization and grain growth of a cold-worked metal like brass as a function of temperature.

5.3 What is the difference between slip and twinning?

5.4 How does recrystallization following cold work differ from recrystallization caused by an allotropic change as in steel?

5.5 How can you refine grain size in pure copper and in 0.8 percent C steel?

5.6 Describe how to seed and grow single crystals of molybdenum of definite orientations.

5.7 An annealed pure FCC crystal will undergo slip when the resolved shear stress (on the slip plane) is 48 psi.

(a) Compute the theoretical stress for shear if no dislocations are present and isotropy obtains.

(b) What is the ratio of theoretical to observed shear stress?

5.8 When the dislocation structure of an FCC metal is resolved in a transmission electron microscope, dislocations on slip systems other than $\{111\}\ \langle 10\bar{1}\rangle$ are observed.

(a) Why do these dislocations not produce slip during plastic deformation?

(b) How can these dislocations contribute to strain-hardening of a crystal?

5.9 A zinc single crystal is oriented with the normal to the basal plane making an angle of 60° with the tensile axis, and the three slip directions making angles of 38°, 45° and 84° with the tensile axis. If plastic deformation is first observed at a tensile stress of 330 psi, calculate the critical resolved shear stress for zinc.

5.10 If the critical resolved shear stress for yielding in aluminum is 35 psi, calculate the tensile stress required to cause yielding when the tensile axis is [001].

5.11 Along what crystallographic direction in an FCC crystal must the tensile axis be oriented in order to produce multiple slip on (a) four slip systems, (b) six slip systems, (c) eight slip systems. (*Hint:* Consider the geometry of the octahedron of $\{111\}$ planes in a cube consisting of eight adjacent unit cells.)

5.12 A zinc single crystal is loaded in compression in a direction making an angle of 7° with the basal plane. The basal plane and the twin plane normals are coplanar with the compression axis. The critical resolved shear stress for slip is 130 psi. Assuming that there is a critical resolved shear stress for twinning of 450 psi, determine the value of compressive stress at which initial yielding occurs and state whether the initial yielding is by slip or twinning.

*Given:* (a) $c/a$ ratio $= 1.856$.

(b) Slip direction $[11\bar{2}0]$ is 31° from the compression axis.

(c) Twinning direction is $[10\bar{1}\bar{1}]$ on $(10\bar{1}2)$ plane.

5.13 Titanium, an HCP metal with $c/a$ ratio less than the ideal value of 1.632, may slip on either the (0001) or $\{10\bar{1}0\}$ planes in the $\langle 11\bar{2}0\rangle$ direction.

(a) Sketch a hexagonal prism; identify the (0001) and $\{10\bar{1}0\}$ planes, and the common slip direction, $\langle 11\bar{2}0\rangle$.

(b) For a single crystal of Ti, oriented for basal plane slip, the stress-strain curve appears more like that of FCC-Cu than that of HCP-Mg. Explain.

5.14 Each partial dislocation of a dissociated pure screw dislocation in an FCC metal possesses edge character as well as screw character. Explain why the cross-slip process requires stress-aided thermal activation to operate and therefore is temperature sensitive.

5.15 An FCC single crystal, oriented for single slip, is tested to a high stress (into the parabolic hardening region) at a low temperature, unloaded, and then tested further at room temperature. A large strain occurs at a stress much lower than predicted for the change in temperature alone. This phenomenon is known as *work-softening*.

(a) Explain work-softening in terms of thermally activated cross-slip.

(b) What difference would be noted in surface slip-line appearance before and after work-softening?

5.16 In FCC crystals, an $a/2[10\bar{1}]$ dislocation may dissociate:

$$\frac{a}{2}[10\bar{1}] \rightarrow \frac{a}{6}[11\bar{2}] + \frac{a}{6}[2\bar{1}\bar{1}]$$

The motion of the pair of partial dislocations results in slip. $a/6[11\bar{2}]$ is also the twin vector in FCC. On what plane does twinning occur in FCC? Show that twinning in FCC results in a shear strain $\sqrt{2}/2$.

5.17 Cd and Be, as well as some other HCP metals, twin on the $(10\bar{1}2)$ plane in the $[10\bar{1}\bar{1}]$ direction.

(a) Derive an equation which gives the twinning shear strain for $(10\bar{1}2)$ $[10\bar{1}\bar{1}]$ twinning as a function of $c/a$ ratio only.

(b) Calculate the twinning shear strain for Cd($c/a$ = 1,886) and Be ($c/a$ = 1.568) and also for a hypothetical metal having a $c/a$ ratio = $\sqrt{3}$.

CHAPTER SIX

# Plastic Deformation

The presence of grain boundaries in polycrystalline materials strongly affects their stress-strain behavior. Certain general features of this behavior can nevertheless be deduced in principle from the behavior of single crystals. At elevated temperature, the deformation of single or polycrystalline material depends on strain rate and temperature.

## 6.1 INTRODUCTION

In Chapter 5, which dealt in large part with the behavior of single crystals, we considered the micromechanisms by which plastic deformation occurs. In single crystals macroscopic behavior can be associated with the microscopic details of deformation, for geometric constraints to dislocation motion are limited to those present within the crystal. Grain boundaries in polycrystalline materials introduce new constraints to dislocation motion. If integrity of the grain boundaries is to be preserved, each grain must deform in a manner compatible with the deformation of all its immediate neighbors. Since the constraints differ from grain to grain, the observed macroscopic behavior of a polycrystalline specimen is the result of a complex give-and-take between a large number of grains. Each grain attempts to deform in its own particular way, but it is forced to compromise by its neighbors. On this account the macroscopic behavior of a polycrystalline metal can, in general, be only qualitatively associated with the behavior of single crystals of the same metal. In this chapter we shall consider the role of grain boundaries, temperature, strain rate, and several other factors which influence the observed behavior of polycrystals.

## 6.2 GRAIN BOUNDARIES

Before comparing the deformation of polycrystal and single crystal materials it is worth while reviewing the description of a grain boundary given in Volumes I and II. Grain boundaries may be formed in a metal after solidification or after cold work and recrystallization. If the mismatch in crystallographic orientation is less than about 10°, the grain boundary is semicoherent and is called a *low angle boundary.* The atoms in such boundaries exhibit a large measure of coherency with the grains themselves, but this coherency is not complete. Low angle grain boundaries may be delineated by dislocation walls. A small angle boundary is consequently often portrayed as an array of edge dislocations, as has been shown in Volume I, Figure 4.11, or as an array of screw dislocations, as shown in Volume I, Figure 4.17. In the former case the low angle boundary is called a *tilt* boundary; in the latter case it is called a *twist* boundary. Actually low angle boundaries are in general part tilt and part twist boundaries.

When the mismatch between crystallographic directions in adjacent grains is greater than about 10°, as is the case in most solidified or cold-worked and recrystallized materials, the surface between the grains is called a large angle boundary. In such a boundary there is little lattice continuity between the grains, even though the boundary itself is only a few atoms thick.

Grain boundaries at low temperatures are usually stronger than the material in the grains themselves. Thus fracture in most metals at low temperature is *transcrystalline.* This signifies that the fracture goes through the interior of the grains rather than along the grain boundaries. At low temperatures polycrystalline metals can withstand stresses on the order of E/100 before failure. In polycrystalline nonmetallics, however, the grain boundaries are weaker than the grains, even at low temperatures, because of poor bonding in the boundary. This *intergranular* weakness is similar to that found in metals at elevated temperatures.

In metals, the strength of grain boundaries is affected by alloying elements and impurities, both of which tend to segregate there. Such segregations can embrittle the material at low temperatures and in some cases even at high temperatures (*hot-shortness*). Segregation at grain boundaries may also accelerate local corrosion (e.g., *sensitization* of stainless steel).

The barrier to slip across a grain boundary has been found experimentally to depend on grain size. If $\sigma_y$ is the yield stress, $d$ the grain diameter, $k$ an empirical constant, and $\sigma_i$ the "friction" stress characteristic of the metal, it has been found that

$$\sigma_y = \sigma_i + kd^{-1/2} \tag{6.1}$$

In this equation, known as the Petch equation, the "friction" stress is thought to be a measure of the intrinsic resistance of the material to the motion of a dislocation. It is temperature dependent.

The effect of grain size on strain-hardening rate as discussed in the next section is valid only for strains of a few percent. Thereafter, the strain-hardening rate is almost independent of grain size.

## 6.3 STRAIN-HARDENING

Tensile stress-strain curves for various single and polycrystalline specimens of Cu are shown in Figure 6.1. It is evident here that polycrystalline copper is stronger than a single crystal of cop-

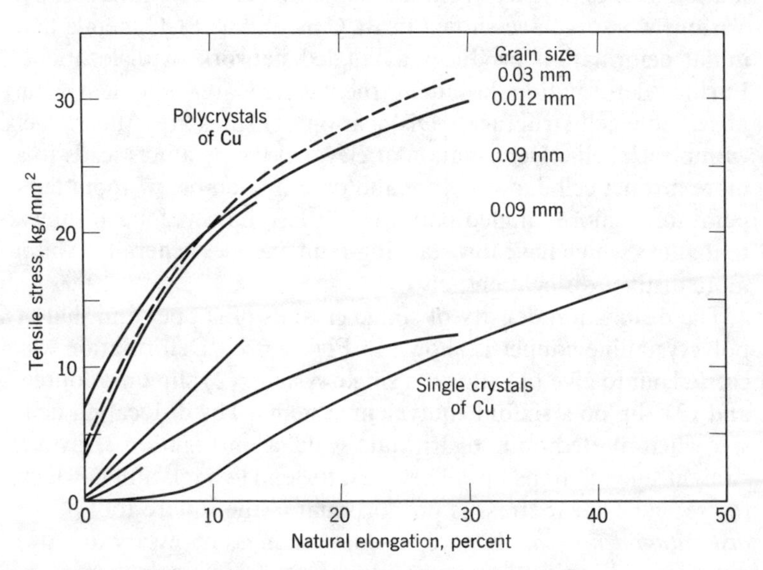


Figure 6.1 Tensile stress-strain curves for single crystals and polycrystalline copper (from D. McLean, *Mechanical Properties of Metals,* John Wiley and Sons, New York, 1962).

per which exhibits single slip. When a single crystal of copper is, however, oriented in the tension test so that six slip systems operate, the curve is hardly different than that for polycrystalline copper. In all the curves of Figure 6.1, it is evident that the stress necessary for continuous flow increases with plastic deformation. In the curve on the right, for the crystal oriented to produce single slip, initial gliding is of the easy kind of Stage 1 in which the strain-hardening rate is very low. In Stage 2 the stress necessary for flow increases nearly linearly with strain. The rate of work-hardening decreases in Stage 3 as flow progresses. In the case of six slip systems operating during tensile extension, the slope of the curve is steeper to begin with. Strain-hardening is much stronger, yet the slope of the curve is steeper from the start. The slope of the curve (rate of hardening) decreases as flow progresses. The latter curve is very much like both curves for polycrystalline material. Multiple slip occurs in all the grains of polycrystalline copper in order that continuity of deformation across grain boundaries occurs.

Our present experimental knowledge of the mechanism of deformation comes largely from transmission electron microscopy. Various workers have shown in BCC as well as FCC metals that initial deformation produces a tangled network of dislocations. Further deformation broadens the tangled area and leaves an observable cell structure. Dislocation densities are much lower within each cell. Deformation at elevated temperatures leads to a more distinct cellular structure and deformation below room temperature to more tangled patterns. It has, however, been shown that alloys which have low stacking-fault energies generally exhibit more distinct dislocation networks.

The dislocation density of single crystals of copper and that of polycrystalline copper is shown in Figure 6.2. Deformation was carried out to give (1) slip on a single system, (2) slip on a double, and (3) slip on a sixfold equivalent system. The dislocation density when plotted on a logarithmic scale, as in Figure 6.2, gives a straight-line relationship. These results tend to a relation in which the resolved shear stress is proportional to the square root of the *dislocation density* $\rho$. If $\tau_0$ represents the stress necessary to move a dislocation in the absence of interfering dislocations, then

$$\tau = \tau_0 + AGb\sqrt{\rho} \tag{6.2}$$

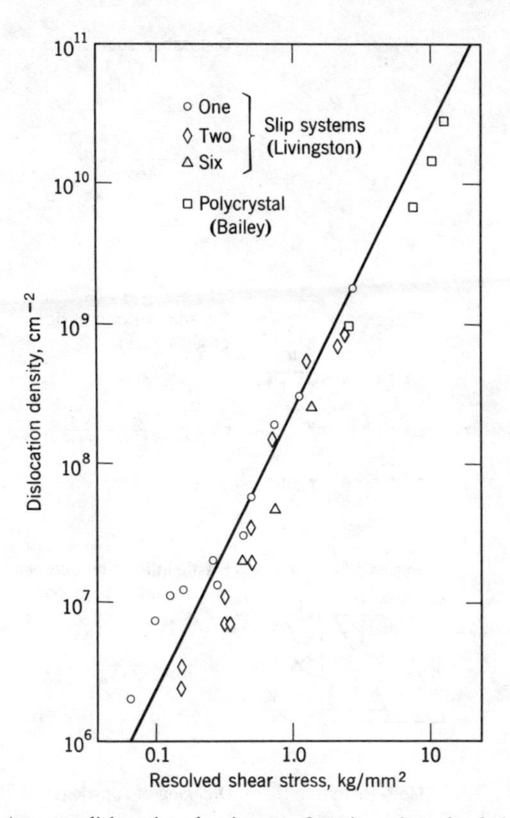


Figure 6.2 Average dislocation density as a function of resolved shear stress for copper; one slip system; two slip systems; six slip systems according to Livingston and polycrystalline Cu according to Bailey (from H. Wiedersich, *Journal of Metals,* May 1964).

where $G$ is the shear modulus, $b$ the magnitude of the Burgers vector, and $A$ is a constant which may take values of 0.3 to 0.6 for various FCC, BCC single and polycrystalline metals as well as single crystals of sodium chloride. It is therefore concluded that strain-hardening is related in some unique way with an increase in dislocation density.

Five different mechanisms have been advanced at various times to describe strain- or work-hardening. These have been arranged schematically as shown in Figure 6.3. The top of Figure 6.3 stems from the original work of Taylor in which it was assumed that

Taylor: Dislocation superlattice

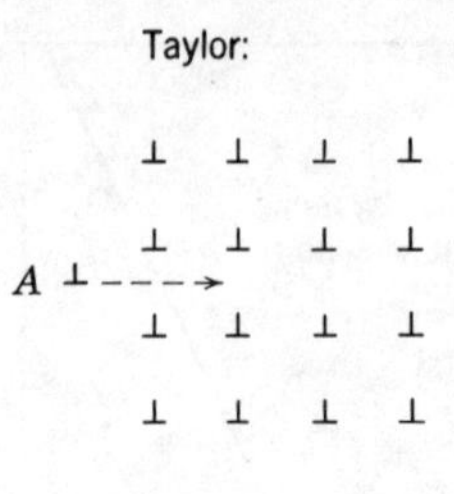


$$\frac{\tau_G}{G} = \frac{1}{2\pi K} b\sqrt{\rho}$$

Seeger: Superlattice of pile-ups or glide zones

B

$n$, $s$

$$\frac{\tau_G}{G} = \frac{\sqrt{n}}{2\pi K} b\sqrt{\rho}$$

Basinski: Elastic interaction between intersecting dislocations

C

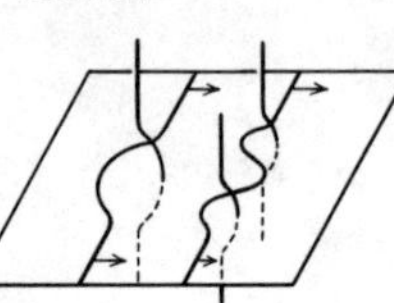

$$\frac{\tau_G}{G} = \frac{1}{2\pi K_B} b\sqrt{\rho_f}$$

Mott-Hirsch: Dragging of superjogs

D

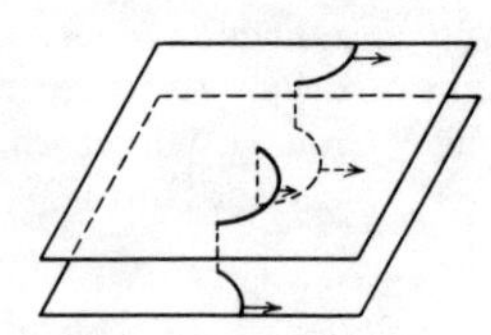

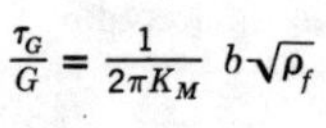

$$\frac{\tau_G}{G} = \frac{1}{2\pi K_M} b\sqrt{\rho_f}$$

Kuhlmann-Wilsdorf: Bowing-out between network points in dislocation tangles

E

$$\frac{\tau_G}{G} = \frac{1}{\pi K_k} b\sqrt{\rho_f}$$

Figure 6.3 Comparison of five models of strain-hardening according to Wiedersich; $\rho$ signifies dislocation density parallel to slip dislocation and $\rho_f$ the density of forest dislocations (from H. Wiedersich, *Journal of Metals,* May 1964).

work-hardening arose from the elastic interactions between dislocations. Equation $A$ in the figure assumes that the stress necessary to force a dislocation through an array must be as large as the average value of the internal stress caused by all dislocations. The equation and symbols in $B$ are supposed to portray a model in which a *superlattice* of pile-ups is bordered by *sessile* dislocations. Here the *back stresses* of the piled-up dislocation groups are supposed to block the movement of newly created dislocations. Basinski pictures a process involving elastic interactions between dislocations moving in a slip plane. The dislocations intersecting the slip plane illustrate *forest dislocations.* According to Mott and Hirsch the flow stress depends on how *jogs* hinder the motion of screw dislocations. The number of jogs which form according to the latter view is related to the density of forest dislocations. The final picture, $E$, is based on the supposition that the flow stress must be large enough to bow the dislocations out of dislocation tangles. Since the $A$'s in the formulae given are the same and $K$ is in most cases unity, the formulae are all more or less alike.

To relate strain to dislocation density in Equation 6.2 requires an equation of the form

$$d\epsilon = b\pi R^2(\epsilon)\, dN \tag{6.3}$$

where it is imagined that dislocation rings are emitted by sources from which they spread out to a radius distance $R$ before getting stuck. Then $dN$ is the number of new loops formed per unit volume during $d\epsilon$ and $R$ is a function of the accumulated strain $\epsilon$. The total periphery of the newly created dislocation loops is taken to be the same as the increase in dislocation density or

$$d\rho = 2\pi R(\epsilon)\, dN \tag{6.4}$$

Combining 6.3 and 6.4 gives

$$d\rho = \frac{2}{bR(\epsilon)}\, d\epsilon \tag{6.5}$$

If $R$ is taken as independent of strain, a parabolic stress-strain curve is obtained as in FCC polycrystals. If $R(\epsilon)$ is taken as inversely proportional to the strain (Stage 2) in FCC single crystals, a linear relation relates stress and strain. The function $R(\epsilon)$ has not as yet been adequately derived from the proper theoretical models.

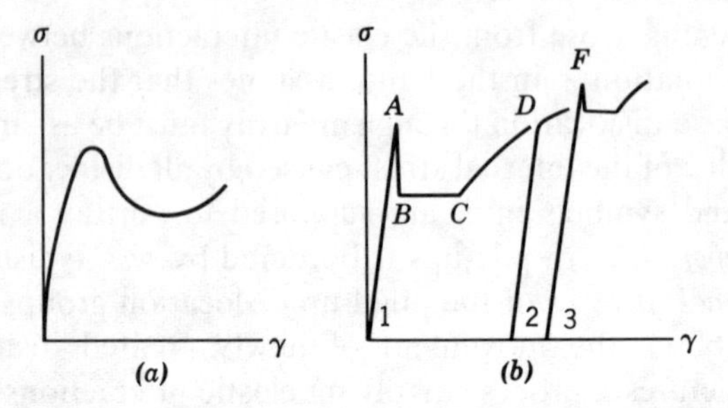


Figure 6.4 Yield drops: (*a*) due to increase of dislocations in Ge crystal; (*b*) due to impurity-pinning of dislocations.

## 6.4 STRAIN AGING

*Etch pits* are useful for locating dislocations. Etch pit studies can also be used to measure dislocation velocity. From there it is found that edge dislocations move faster than screw dislocations. At high stresses the speed of motion may approach that of sound. If the applied stress is $\sigma$ and $\sigma_y$ is the stress per unit speed, then the speed is given by

$$v = \left(\frac{\sigma}{\sigma_y}\right)^n \tag{6.6}$$

where $0 < n < 40$. In the ideal case $n = 0$. Hard crystals such as germanium at high temperatures and low stresses have a value of about $n \simeq 2$ (at 935°C), but in most others $n > 10$.

If a Ge crystal is rapidly strained, the stress is initially high, as shown in Figure 6.4*a*. The small number of dislocations initially present must now move quickly. If the dislocations multiply, the individual ones need not move as quickly; hence the yield stress falls, giving a characteristic yield drop as shown. Such yield drops are found in metals as well as in nonmetallics.

It is possible to pin dislocations by foreign atoms as in the so-called aging of mild steel. Here interstitial carbon and nitrogen atoms can migrate to the space provided by dislocations. Segregations of such impurity atoms along dislocations are called "*Cottrell atmospheres*" and precede precipitation as true carbides or nitrides.

*Strain aging* in mild steel may be represented as in Figure 6.4*b*. The stress drops sharply from the upper yield point $A$ to a lower yield point $B$. $A$ yield elongation $BC$ then takes place. During this time a Lüders band forms at constant stress. Thereafter plastic deformation takes place in normal fashion. If we now interrupt the test, unload and then reload shortly thereafter, as shown in Figure 6.4*b* by the curve starting at $D$ and reload, no yield drop is observed; for enough dislocations that are free to glide are present. On the other hand, if we let the specimen rest for some time, reloading gives another curve starting at $F$ similar to the initial curve but with a higher yield stress. Since such strain aging depends on the rate of diffusion of carbon and nitrogen in BCC iron, it takes only a few minutes at 100°C but a few days at room temperature. Lüders bands produced by strain aging in deep drawing are called *stretcher strain* markings. Since they disfigure the steel they are purposely prevented by lightly prerolling the sheet before deep drawing.

## 6.5 TEMPERATURE DEPENDENCE

The effect of grain size on the yield stress of well-annealed steel is not highly temperature sensitive, whereas the effect of *unpinning* dislocations from their locking atmospheres is quite temperature sensitive. Thus the value of $k$ in the Petch equation (6.1) varies as shown in Figure 6.5.

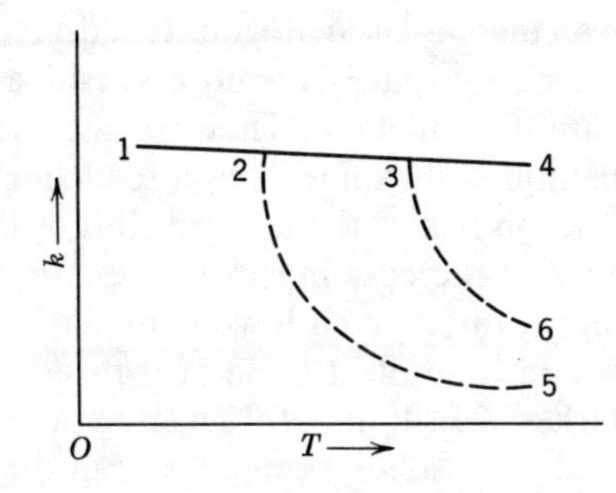


Figure 6.5 Variation of $k$ of Petch equation with temperature.

Here line 1-2-3-4 represents heavily aged material; less severely aged samples continue along curve 3 to 6 and lightly aged samples along 2 to 5 as the temperature is raised. Along the curves 1 to 2 or 1 to 3 new grain boundaries are formed. Along curves 2 to 5 and 3 to 6, the high temperature parts, unpinning occurs. Now, if actual precipitates of carbides or nitrides form around dislocations, the unpinning process ceases to be temperature dependent. Thus the dislocations can be torn away from the precipitates. If the precipitate coarsens, the yield stress is further reduced. These effects have also been noted in HCP or BCC metals where nitrogen is the interstitial solute. In regard to iron it is well to point out that no yield point is observed above 400°C. Above this temperature the Cottrell atmospheres are readily dispersed by thermal vibrations.

The strain-hardening rate of HCP metal sheet and rod can be high at room temperature due to the limited number of slip systems available. With increasing temperature more independent slip systems become available, and grain boundary sliding becomes more prevalent.

The change in the rate of work-hardening as the temperature is lowered is usually much smaller in BCC than in FCC metals. The total change in hardness at a given strain can, however, be much larger in BCC than in FCC metals.

## 6.6 STRAIN RATE

The tensile stress-strain curve can be raised by an increase in strain rate; such an increase is analogous to a decrease in temperature. BCC metals are usually far more sensitive to changes in strain rate than are FCC metals. They are also more sensitive to changes in temperature. In some cases a reduction in ductility or total elongation accompanies the upward shift of the stress strain curve produced by an increase in strain rate. In other materials an increase in strain rate, which lowers ductility in certain temperature ranges, does not affect them or it raises them in others. Frequently a marked sensitivity of flow stress and strain rate, as in BCC metals and ionic solids, portends a transition from ductile to brittle behavior at some low or intermediate temperature.

## 6.7 STRAIN RATE AND TEMPERATURE

The mechanical behavior described in the last two sections indicates that both increasing the strain rate and decreasing the temperature raises the level of the tensile stress-strain curve. At low temperatures and high strain rates some BCC metals, such as plain carbon steels, molybdenum, and tungsten, undergo such a large increase in initial yield strength that cleavage may occur before the generation of appreciable plastic flow. FCC metals, such as copper and aluminum, tend to exhibit a much less drastic increase in strength but little if any reduction in ductility. They also show no abrupt decrease in impact energy absorption at low temperatures. For this reason aluminum alloys and *austenitic* iron-nickel alloys are useful in low temperature applications.

The combination of high strain rate and elevated temperature is of interest because the lowering of the stress-strain curve at elevated temperatures may be compensated by increasing the strain rate. Although this compensation occurs only while the high strain rate persists, it may still be important for those applications where the required life of the part can be short. At high temperatures strength decreases for various reasons. The work-hardening rate decreases as the temperature increases, for recovery processes become more active, and it is impossible to maintain a "cold-worked" structure. In addition, as the temperature is further increased, recrystallization and grain growth may occur, eliminating the benefit which accrues from a fine grain size. If, however, plastic flow occurs more rapidly than diffusion can eliminate its effects, the metal can then be used under rapid loading at high stress levels, even at high temperatures. Although the static stress-strain curve may drop very rapidly with increasing temperature, the percentage increase in strength due to impact loading is much greater at elevated than at low temperatures. For a number of different materials strength and strain rate data are given in Figure 6.6. Here a rapid increase in the effect of high strain rate occurs at temperatures of the order of 40 to 50 percent of the absolute melting point of the metal. This is the temperature at which recrystallization proceeds rapidly in most metals. The curves for the light metals, aluminum and magnesium, lie closer to the right-hand dashed curve, and metals in which structural changes occur

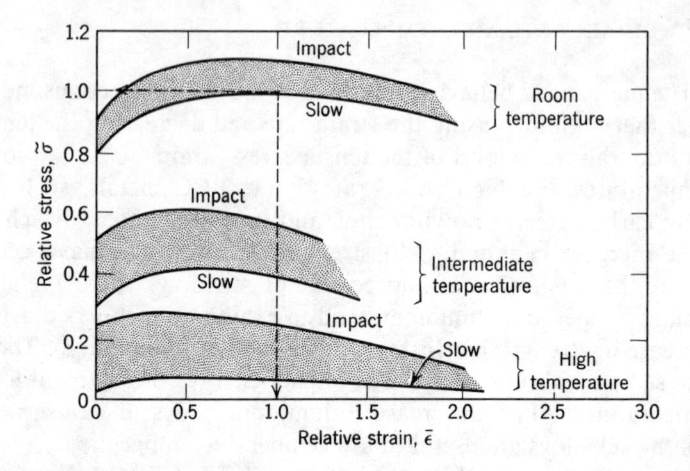


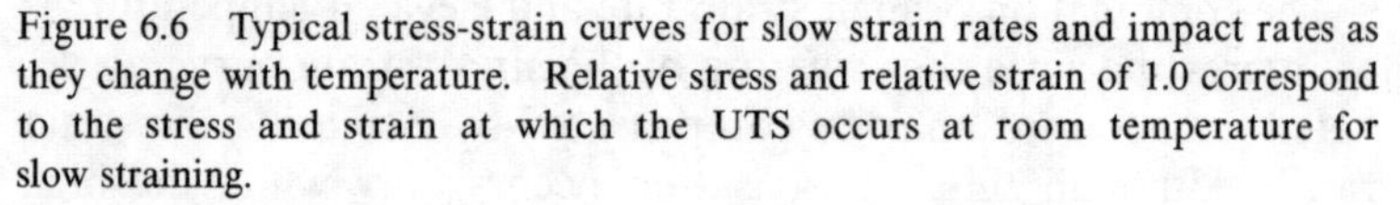
Figure 6.6 Typical stress-strain curves for slow strain rates and impact rates as they change with temperature. Relative stress and relative strain of 1.0 correspond to the stress and strain at which the UTS occurs at room temperature for slow straining.

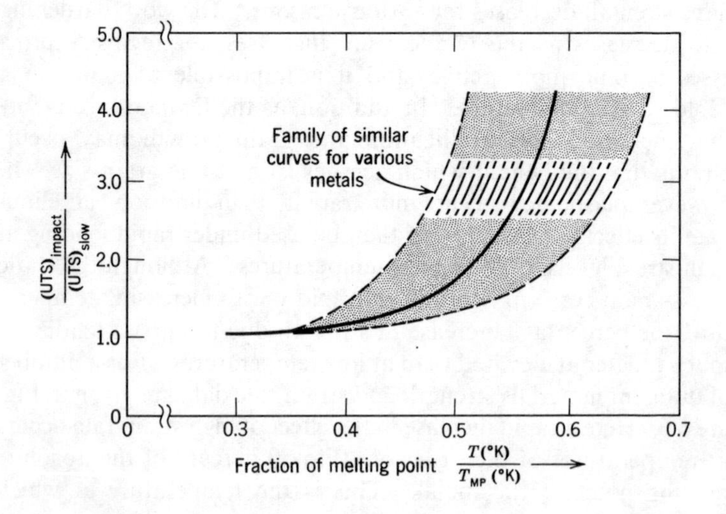


Figure 6.7 Ratio of UTS for impact and slow strain rates as a function of temperature.

readily at modestly low temperatures lie closer to the left-hand dashed curve. The strength of metals at elevated temperature is greater under impact strain rates as compared to slow strain rates. The high temperature impact stress-strain curve lies considerably below the room temperature slow strain rate curve, as shown in Figure 6.7.

## 6.8 CREEP

The question of *creep* or the amount of deformation that might be expected from a given material under a constant stress as a function of time and temperature is of great practical importance. Creep is usually defined as the continuing or *time-dependent deformation* of a material under a constant stress or load. In contrast to constant strain-rate deformation, the amount and rate of straining during creep are established by the material itself under the imposed stress and temperature conditions.

Because crystalline materials work-harden in the process of deformation, continuing deformation under constant stress implies some recovery. The *creep rate* (strain rate) then results from a balance of simultaneous work-hardening and recovery processes. The two significant stress-aided and thermally activated recovery processes in crystals, namely, cross-slip and dislocation climb, have already been discussed in Chapters 4 and 5. At lower temperatures, the cross-slip of screw dislocations is the only process by which obstacles in the slip plane can be bypassed or dislocations annihilated. At temperatures above $\frac{1}{2}T_M$ (absolute melting point), the vacancy concentration is higher and the diffusion rate more rapid. Recovery by dislocation climb then becomes of prime importance.

*Creep* curves for a pure polycrystalline metal at temperatures above and below $\frac{1}{2}T_M$ are drawn schematically in Figure 6.8. For both creep curves note the initial plastic strain upon application of the load, followed by an initial period of creep at a rapidly decreasing strain rate (*primary* creep). At the lower temperature the creep rate continues to decrease with time unless the stress is of the order of the ultimate tensile strength. In terms of recovery,

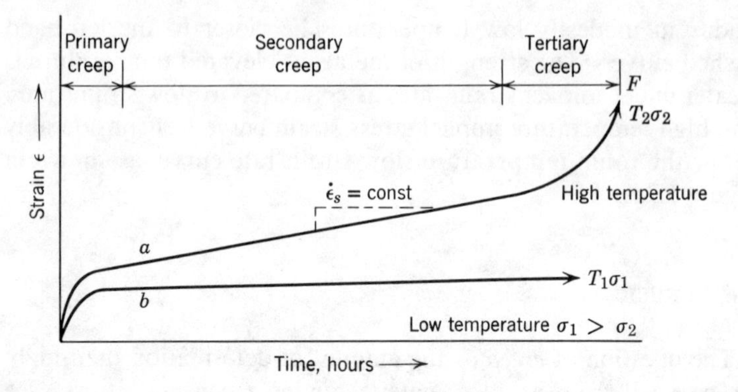


Figure 6.8 (*a*) The three stages of the strain-time relation in high-temperature creep. (*b*) The same relation at low temperature.

the number of points at which cross-slip can aid in the glide or annihilation of dislocations is eventually exhausted. The material has work-hardened to support the applied stress, and the creep rate becomes negligibly small. At the higher temperature, primary creep is followed by a period of extension at constant strain rate; this is called *secondary* creep. During secondary creep, the rate of work-hardening is exactly balanced by the recovery rate. Under these conditions the creep resistance of the structure is greatest. *Tertiary* creep can be due to necking of the specimen as well as to the accumulation of voids (particularly at grain boundaries). Grain boundary sliding (Chapter 5) also makes a significant, yet secondary contribution to high temperature creep straining. In any case, fracture always terminates tertiary creep at high temperatures.

The high-temperature creep process is essentially limited by the rate of climb of blocked dislocations from their slip plane. Climb is controlled by the rate of diffusion of vacancies to or from dislocation jogs under the action of the local stress fields. The *Dorn-Weertman* relation developed from these concepts is usually written

$$\dot{\epsilon} = A\sigma^n e^{-Q_{SD}/RT} \tag{6.7}$$

where $n \simeq 5$, the creep strain rate $\dot{\epsilon}$ during secondary creep is expressed as a function of the stress, $\sigma$, and the activation energy for self-diffusion is $Q_{SD}$. This equation agrees with experimental data for pure metals and dilute alloys in the practical creep stress range.

Neither the Dorn-Weertman relation nor other more empirical parameters used in interpolating or extrapolating creep data predict a functional form of the creep curve. Instead, they are normally used in prediction of time to any given strain or fracture. A modified relationship of the form

$$te^{-Q/RT} = f(\sigma) \tag{6.8}$$

where $t$ is the time for a given strain, can also be used.

The most commonly applied creep relation, called the Larson-Miller parameter, is both simple and applicable to a wide variety of materials, both single phase and dispersion-hardened. The Larson-Miller parameter sets time for a given strain or fracture according to

$$T(C + \log t) = f(\sigma). \tag{6.9}$$

In this relation the temperature $T$ is given in degrees Rankine (°F + 460) and the time $t$ in hours; the constant $C \simeq 20$ for a large number of alloys. A typical Larson-Miller plot for the creep fracture of a dispersion hardened nickel base super-alloy, M252, is shown in Figure 6.9.

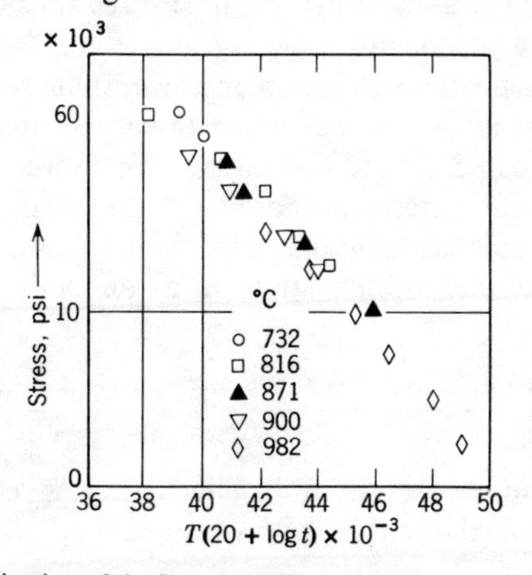


Figure 6.9 Application of the Larson-Miller parameter to the fracture force of Ni alloy M252. Data for 732°C falls at the top of the band, data for 871°C in the middle, and data for 982°C at the bottom (from A. J. Kennedy, *Processes of Creep and Fatigue in Metals,* John Wiley and Sons, New York, 1963).

Most high-temperature creep-resistant alloys in use today are precipitation-hardened. The effects of a fine uniform dispersion of precipitate particles in limiting creep are in part associated with additional barriers to slip and grain boundary sliding that the particles impose. Such dispersed particles serve furthermore to restrict recovery. Although this process is not completely understood, it is clear that high dislocation densities can be maintained in these materials even close to the melting temperature.

## 6.9 COMBINED STRESSES

Although the yielding and work-hardening characteristics of materials are usually determined in uniaxial tension, the practical use of these metals often involves complex *stress states.* Any stress state, however complex, involving both normal and shear stresses in three dimensions can be resolved into a new equivalent set of three mutually perpendicular normal stresses only, called *principal* stresses. If the three principal stresses are $\sigma_1$, $\sigma_2$, $\sigma_3$, then the shear stresses caused by any pair are $(\sigma_1 - \sigma_2)/2$, $(\sigma_2 - \sigma_3)/2$, $(\sigma_3 - \sigma_1)/2$. It is then clear that in the case of hydrostatic stresses ($\sigma_1 = \sigma_2 = \sigma_3$) shear stresses do not exist in the material. Thus dislocations cannot move, and the material cannot distort plastically with increasing *hydrostatic stress.* It is only when there is an imbalance in the three principal stresses that shear stresses can exist and dislocations can move.

Of the several criteria proposed for the yielding of a metal, probably the most satisfactory is the Von Mises criterion, which postulates that the *effective* stress in a randomly oriented polycrystal is

$$\bar{\sigma} = \left[ \frac{(\sigma_1 - \sigma_2)^2 + (\sigma_2 - \sigma_3)^2 + (\sigma_3 - \sigma_1)^2}{2} \right]^{1/2} \tag{6.10}$$

and that yielding occurs when the effective stress reaches the same value as the uniaxial tensile yield stress, $\sigma_y$:

$$\sigma_y = \left[ \frac{(\sigma_1 - \sigma_2)^2 + (\sigma_2 - \sigma_3)^2 + (\sigma_3 - \sigma_1)^2}{2} \right]^{1/2} \tag{6.11}$$

Note that the effective stress (and hence yielding) depends only on the shear stresses $(\sigma_1 - \sigma_2)/2$, $(\sigma_2 - \sigma_3)/2$, $(\sigma_3 - \sigma_1)/2$; this is consistent with the fact that on a fine scale, plastic flow occurs by the shear mechanisms, slip and twinning.

Now, deformation under combined stresses causes work-hardening, as does deformation under uniaxial stress. The value of effective stress, $\bar{\sigma}$, required to continue plastic deformation increases with the effective strain, $\bar{\epsilon}$;

$$\bar{\epsilon} = \frac{\sqrt{2}}{3}[(\epsilon_1 - \epsilon_2)^2 + (\epsilon_2 - \epsilon_3)^2 + (\epsilon_3 - \epsilon_1)^2]^{1/2} \tag{6.12}$$

Since the effective stress, $\bar{\sigma}$, and the effective strain, $\bar{\epsilon}$, reduce to true tensile stress, $\sigma$, and true tensile strain, $\epsilon$, for simple tension, the tension test may be used to determine the dependence of $\bar{\sigma}$ on $\bar{\epsilon}$. If the experimental curve for uniaxial tension can be represented adequately by the power-law equation

$$\sigma = K\epsilon^n \tag{6.13}$$

then the general, three-dimensional case can be represented by

$$\bar{\sigma} = K\bar{\epsilon}^n \tag{6.14}$$

for Equation 6.14 reduces to Equation 6.13 when $\bar{\sigma}$ is computed for uniaxial tension ($\sigma_1$ = applied tensile stress, $\sigma_2 = \sigma_3 = 0$), and when $\bar{\epsilon}$ is similarly computed for the uniaxial tension test ($\epsilon_1$ = axial strain and $\epsilon_2 = \epsilon_3 = -\frac{1}{2}\epsilon_1$, where Poisson's ratio, $\nu$, is taken as $\frac{1}{2}$ because the strains are plastic).

## 6.10 PLASTIC CONSTRAINT

Even when the applied stress is merely uniaxial, complex stresses often occur because of the geometry of the deforming material or because of local differences in elastic or plastic behavior within the material.

Consider a thin wafer of soft material bonded between two hard materials, as, for example, the soldered joint illustrated in Figure 6.10. If the softer material is to elongate under a tensile stress $\sigma_1$ applied across the joint, it must also contract laterally. However, the adjoining material to which it is bonded resists the lateral contraction, effectively subjecting the thin wafer to tensile stresses, $\sigma_r$,

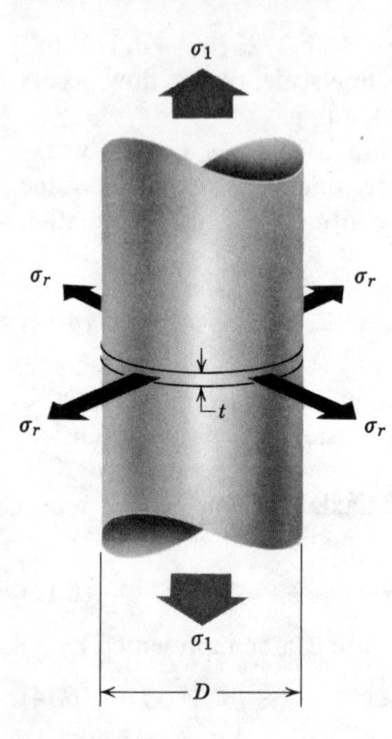


Figure 6.10 Generation of transverse stresses in a soldered joint subjected to uniaxial tension.

Figure 6.11 Geometry of the necked region of a tensile specimen.

directed radially outward. Now, as far as the thin wafer is concerned, $\sigma_2$ and $\sigma_3$ are not zero and the Von Mises yielding criterion, Equation 6.11, will not be satisfied unless $\sigma_1$ is much greater than the normal unconstrained yield strength, $\sigma_y$. The actual value of $\sigma_1$ necessary to cause yielding depends on the geometry of the wafer: if the ratio $t/D$ is of the order of 1 to 2, then the wafer is similar in geometry to a standard tensile specimen and can yield at the center, independently of the presence of the hard interfaces. However, as the $t/D$ ratio becomes smaller, the wafer is not capable of necking in the same way as a standard tensile specimen without generating interface constraints; thus the value of $\sigma_1$ necessary to cause yielding increases as $t/D$ decreases. A similar effect is observed in the compression of discs with a small $t/D$ ratio.

An analogous situation occurs when a neck has formed in a cylindrical tensile specimen. The diameter of the specimen is a minimum at the center of the neck; on either side of the neck, therefore, where the diameter is greater, the stress is lower. Since the stress is lower outside the neck, the strain is smaller; thus an outward radial stress is generated at the minimum section. The constraint increases with the severity of the neck (expressed as $a/r$—see Figure 6.11—where $a$ is the radius of the minimum cross section and $r$ is the radius of curvature of the neck). Bridgeman has shown that the average ratio of effective stress across the neck to the true tensile stress is given by

$$\frac{\bar{\sigma}}{\sigma} = \frac{1}{[1 + (2r/a)] \ln [1 + (a/2r)]} \tag{6.15}$$

This correction must be applied if tensile data taken after necking starts are to be used to establish effective stress-strain relationships.

## DEFINITIONS

*Creep.* Continuing deformation, with the passage of time, in materials subjected to constant load or stress.

*Viscous.* Descriptive of constant strain rate: $d\epsilon/dt$ = constant.

*Primary Creep.* A transient component of creep for which the creep rate diminishes with time.

*Secondary Creep.* The constant strain-rate component of creep.

*Yield Point.* The stress at which plastic flow initiates in short-time loading.

*Plastic Constraint.* The inhibition of plastic flow resulting from the generation of a multiaxial stress system.

## BIBLIOGRAPHY

INTRODUCTORY READING:

Guy, A. G., *Physical Metallurgy for Engineers,* Addison-Wesley, Reading, Mass., Chapters 6 and 7.

Keyser, C. A., *Materials of Engineering,* Prentice-Hall, Englewood Cliffs, N.J., 1956, Chapters 9 and 10.

Van Vlack, L. H., *Elements of Materials Science,* 2nd ed., Addison-Wesley, Reading, Mass., 1964, Chapters 3, 8, 12, 13.

SUPPLEMENTARY READING:

Cottrell, A. H., *The Mechanical Properties of Matter,* John Wiley and Sons, New York, 1964, Chapters 9, 10, 11.

Dieter, G., *Mechanical Metallurgy,* McGraw-Hill Book Co., New York, 1961, Chapters 4, 5, 6.

Grant, N. J., *Creep and Fracture at Elevated Temperatures—Heat Resistant Alloys,* A.S.M. Cleveland, 1954.

McLean, D., *Mechanical Properties of Metals,* John Wiley and Sons, New York, 1962, Chapters 4, 5, 9, 10.

Smith, C. V., *Properties of Metals at Elevated Temperatures,* McGraw-Hill Book Co., New York, 1950.

ADVANCED READING:

Hill, R., *Mathematical Theory of Plasticity,* Oxford University Press, London, 1950.

Nadai, A., *Theory of Flow and Fracture of Solids,* 2nd ed. McGraw-Hill Book Co., New York, 1950.

Prager, W., *Introduction to Plasticity,* Addison-Wesley, Reading, Mass., 1959.

Wiedersich, H., "Hardening Mechanisms and Theory of Deformation," *Journal of Metals,* May 1964, p. 425.

## PROBLEMS

6.1 (a) If strain energy is associated with every dislocation, should not energy also be associated with a grain boundary?

(b) What are the units for surface energy of a grain boundary?

(c) If free metal surfaces have a value of 1200 to 1800 dynes for FCC metals, what percentage of this would you estimate grain boundaries have?

6.2 (a) When is grain boundary shear possible in metals?

(b) How does it depend on temperature?

(c) How can you relate it to dislocation climb?

6.3 Metals held at high temperatures in inert atmospheres often exhibit grain boundary grooving. If a boundary makes a 90° angle with the surface, what is the dihedral angle at the bottom of the groove? Assume a grain boundary energy of 500 ergs/cm$^2$.

6.4 How could you experimentally reveal grain boundary shear?

6.5 Rods of aluminum, copper, and low-carbon steel are stretched fast enough that the deformation is approximately adiabatic. Estimate the temperature rise in each material at the start of necking. The stress-strain curves for each of these materials can be expressed in the form $\sigma = K\epsilon^n$. The following table gives appropriate numerical values for $K$ and $n$ and also for density, $\rho$, and specific heat, $C_P$:

| Material | $K$(psi) | $n$ | $\rho$(g/cc) | $C_p$(cal/g°C) |
|---|---|---|---|---|
| Al | $5 \times 10^4$ | 0.2 | 2.67 | 0.23 |
| Cu | $4.5 \times 10^4$ | 0.5 | 8.96 | 0.095 |
| Steel | $6.5 \times 10^4$ | 0.3 | 7.87 | 0.12 |

Assume that the stored energy of cold work is small in comparison with the total deformation energy (actually, it is usually only of the order of 2 to 3 percent of the total energy at the point of necking) and that the numerical values of $K$ and $n$ are not materially affected by the temperature rise involved. Recall that necking occurs when $\epsilon = n$. 1 cal = 37 in.-lb.

6.6 (a) Indicate how stress rupture tests are made at elevated temperatures.

(b) Draw a plot showing stress as the ordinate and rupture time in hours as the abscissa.

(c) How are these data applied?

6.7 Draw a curve plotting cohesive energy as ordinate and temperature as abscissa to illustrate the point that grain boundaries are weaker above equicohesive temperature than the grains themselves.

6.8 Plot families of curves of UTS versus strain rate for constant temperature and UTS versus temperature for constant strain rate from the following data for an aluminum alloy:

| | UTS, ksi, at strain rate of | | |
|---|---|---|---|
| $T$, °F | $\dot{\epsilon} = 10^{-3}$ sec$^{-1}$ | $\dot{\epsilon} = 10^{-1}$ sec$^{-1}$ | $\dot{\epsilon} = 10$ sec$^{-1}$ |
| 200 | 62 | 64.5 | 66.5 |
| 400 | 49 | 53 | 57 |
| 600 | 25.5 | 33 | 41 |
| 800 | 8 | 16.5 | 25 |

(a) Take the UTS at $T = 400°$F, $\dot{\epsilon} = 10^{-3}$ sec$^{-1}$ as a reference point: increasing the strain rate by a factor of $10^4$, at constant temperature, is the equivalent of what temperature change at constant strain rate in its effect on the UTS?

(b) Take the UTS at $T = 600°$F, $\dot{\epsilon} = 1$ sec$^{-1}$ as a reference point: increasing the temperature by 60°F at constant strain rate is the equivalent of what change in strain rate at constant temperature in its effect on the UTS?

6.9 For a simple tension test, in the region between initial yielding and the UTS, show that the effective stress and effective strain reduce to the applied axial stress and the resulting axial plastic strain.

6.10 A sample of metal is plastically deformed by application of a stress system given by the principal stresses $\sigma_1$, $\sigma_2$, $\sigma_3$. A hydrostatic stress $p$ is now superposed on the existing stresses; show that the equivalent stress, $\bar{\sigma}$, is not changed and thus that no further deformation occurs.

6.11 Explain why lead in the form of thin gaskets can be used at compressive stresses far in excess of its unconstrained yield strength, $\sigma_y$.

6.12 (a) Plot UTS of cold-rolled steel sheet versus prior strain in percent.

(b) Do the same for yield strength versus prior strain.

(c) How do these curves change after annealing the material?

(d) What happens to yield strength and ultimate tensile strength on annealing? How can you reconcile these at-first-glance anomalous results?

6.13 Consider a rapidly cooled copper ingot.

(a) What is the state of stress at the surface?

(b) What is the state of stress at the interior?

Note that the exterior cools more rapidly than the interior.

6.14 Why is spring material often given a cold rolled or shot-blasted finish?

6.15 What is the state of stress at the surface and at the center of an alloy tool steel that transforms completely to martensite on cooling.

CHAPTER SEVEN

# Fracture

Fracture is the separation of a body into two or more parts. The nature of fracture differs with materials and is often affected by the nature of the applied stress, geometrical features of the sample, and conditions of temperature and strain rate. The differing types of fracture produced in ductile materials and brittle materials, under alternating stress or at high temperatures, arise from differences in the modes of crack nucleation and propagation, which vary for each of these conditions.

## 7.1 INTRODUCTION

*Fracture* is the separation of a body under stress into two or more parts and is usually characterized as either *brittle* or *ductile.* Brittle fracture occurs by the very rapid propagation of a crack after little or no plastic deformation. In crystalline materials brittle fracture usually proceeds along characteristic crystallographic planes called *cleavage* planes, and a brittle fracture surface in a polycrystalline material has a *granular* appearance (Figure 7.1*b*) because of the changes in orientation of these cleavage planes from grain to grain. Brittle fracture can proceed along a grain boundary path instead of along cleavage planes. This is called *intergranular* fracture and may be attributed to embrittling films which have segregated in the grain boundaries. In either case, brittle fracture occurs normal to the maximum applied tensile stress component.

Ductile fracture is fracture occurring after extensive plastic deformation and is characterized by slow crack propagation result-

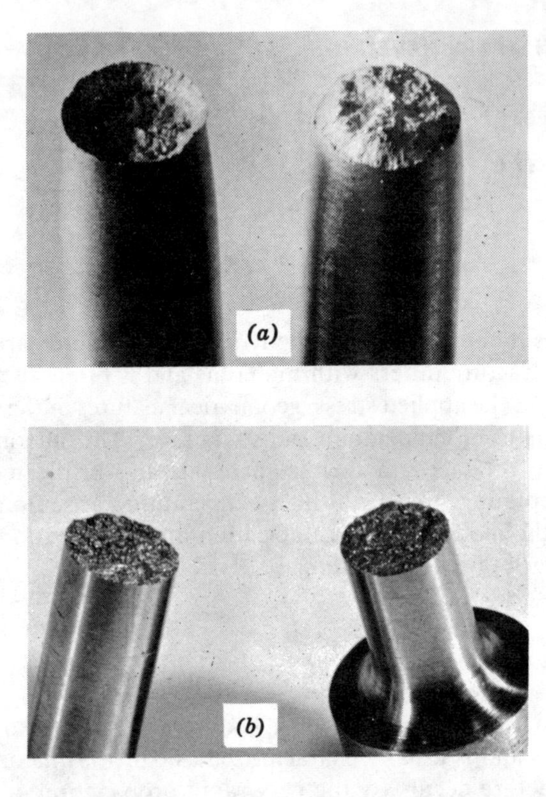


Figure 7.1 (*a*) Ductile (cup and cone) fracture in aluminum (courtesy P. R. LaFrance); (*b*) Brittle fracture in a mild steel (courtesy Behram Kapadia).

ing from the formation and coalescence of voids. A ductile fracture surface has a characteristic dull, fibrous appearance (Figure 7.1*a*).

In BCC transition metals the presence of *notches,* or use at low temperatures, or use under high rates of strain can cause a transition from ductile to brittle fracture. In any such material, ductile fracture is characterized by large energy absorption prior to failure, whereas brittle fracture requires little energy absorption. In using such materials, it is therefore important to avoid situations which induce brittle behavior.

## 7.2 THEORETICAL COHESIVE STRENGTH

An exact calculation of the theoretical fracture strength of a perfect material is a very complicated task. However, using simple models based on known values of heats of sublimation, interatomic forces, and surface energies, estimates can be made. Although each of these approaches differs, the values each predicts for the theoretical cohesive strength are similar and lie between $10^6$ and $10^7$ psi.

These strengths are several orders of magnitude greater than those usually observed. However, when special techniques are employed to produce very perfect materials, strengths approaching the theoretical estimates have been achieved. Extremely fine glass fibers freshly drawn from the melt have strengths approaching the theoretical estimates. However, if these fibers come into contact with hard objects, or even if they are allowed to stand in the atmosphere for a short period of time, their strengths are greatly decreased. This indicates that the strength of the fibers is highly dependent on surface perfection: anything that can give rise to surface irregularities, such as small nicks or cracks, weakens them.

Similar experiments with metal whiskers (fine single-crystal filaments) indicate that both the yield strength and the fracture strength increase as the diameter of the filament decreases. Whiskers can be grown with few dislocations. A decreasing number of dislocations in smaller diameter whiskers results in higher yield and fracture strengths. From such observations, one is led to the conclusion that the difference between the theoretical and observed strengths of materials is caused by structural irregularities.

## 7.3 BRITTLE FRACTURE—GRIFFITH THEORY

Brittle fracture takes place with little or no preceding plastic deformation. It occurs, often at unpredictable levels of stress, by the sudden propagation of a crack. Amorphous materials, such as glass and glassy polymers, are completely brittle; in crystalline materials, however, some plastic deformation precedes brittle fracture.

The first explanation given for the discrepancy between the theoretical strength and actual fracture strength in completely brittle materials was offered by Griffith. He assumed that in a brittle material there are many fine ellipitical cracks, as shown in Figure 7.2, and that at the tip of such cracks there is a strong concentration of stress. The highest stress at the tip of such an elliptical crack can be expressed as

$$\sigma_m \cong 2\sigma\left(\frac{c}{\rho}\right)^{1/2} \tag{7.1}$$

where $\sigma_m$ is the maximum stress at the tip of the crack, $c$ is half the length of an interior crack or the length of a surface crack, $\rho$ is the radius of curvature at the end of the major axis, and $\sigma$ is the applied tensile stress normal to the crack. With such a stress concentration, the theoretical cohesive strength can be attained at this localized area when the body of the material is under a fairly low applied tensile stress.

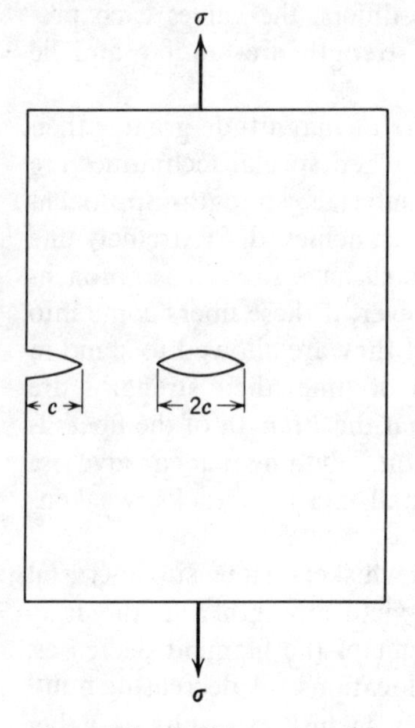


Figure 7.2 Griffith crack model.

When a crack begins propagating, elastic energy is released. However, a certain amount of energy is gained as surface energy due to the creation of new crack surface area. The elastic strain energy per unit thickness released by the spreading of a crack in a thin plate is given by

$$U_E = -\frac{\pi c^2 \sigma^2}{E} \tag{7.2}$$

and the surface energy gained by the creation of the crack is

$$U_s = 4c\gamma \tag{7.3}$$

The crack will propagate and produce brittle fracture when an incremental increase in its length either decreases or does not change the net energy of the system.

$$\frac{\partial U}{\partial c} = \frac{\partial (U_E + U_s)}{\partial c} = -\frac{2\pi c\sigma^2}{E} + 4\gamma = 0 \qquad (7.4)$$

or

$$\sigma = \left(\frac{2\gamma E}{\pi c}\right)^{1/2}$$

The foregoing analysis by Griffith applies to a crack in a thin plate under plane stress. Others have analyzed the behavior of an oblate spheroidal crack in a volume of material. They showed that the original Griffith equation must be modified by a small correction based on Poisson's ratio.

In Equation 7.4, the stress necessary to cause brittle fracture varies inversely with the length of existing cracks. Hence the tensile strength of a completely brittle material is determined by the length of the largest crack existing prior to loading. The relatively low strength of glass is caused by the existence of surface cracks about one to two microns in length. Griffith's experiments with freshly drawn fibers show that when surface cracks are not present, the strengths of the fibers approach the theoretical cohesive strength. The statistical nature of the distribution of cracks in glass fibers was shown by Reinboker, who successively fractured fragments of glass fibers; by thus progressively eliminating the most serious cracks, strength was increased by a factor of two or three.

Orowan made a similar demonstration of the nature of surface cracks by stressing sheets of mica in tension with grips narrower than the width of the sheets. With this apparatus, only the center of the sample was under stress—there was no stress on the edges, where cracks would usually occur. Under these conditions the tensile strength of mica was found to be ten times greater than that observed when its edges were stressed.

The Griffith theory, as it has just been described, is for uniaxial tension. It can be extended to the case of biaxial stress including tension as well as compression. It is then necessary to assume that cracks are randomly distributed and that failure occurs when the stress at any crack tip reaches the value that causes cracks to

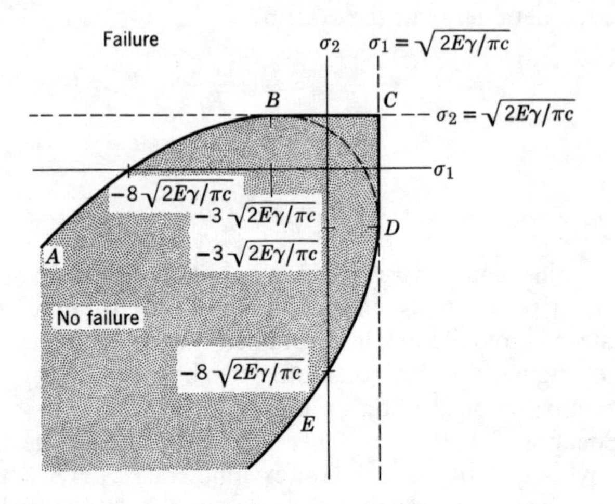


Figure 7.3 The condition for failure according to the biaxial Griffith theory.

propagate under uniaxial tension. If these conditions hold, then the criterion for brittle failure is

$$\sigma_1 = \sqrt{2E\gamma/\pi c} \qquad \text{if } 3\sigma_1 + \sigma_2 > 0 \tag{7.5}$$

$$(\sigma_1 - \sigma_2)^2 + 8\sqrt{(2E\gamma/\pi c)}\,(\sigma_1 + \sigma_2) = 0 \qquad \text{if } 3\sigma_1 + \sigma_2 < 0 \tag{7.6}$$

In these equations $\sigma_1$ and $\sigma_2$ are the two principal stresses, and $\sigma_1$ is algebraically larger than $\sigma_2$. The curves defined by these conditions are shown in Figure 7.3. In this figure, the failure line is shown as the dark line *ABCDE*. Failure will not occur in brittle materials according to the biaxial Griffith theory for those combinations of stress which define the shaded region enclosed by the curve. If either one of the stresses $\sigma_1$ or $\sigma_2$ is zero (i.e., uniaxial compression), it can be seen from this figure that the stress required to cause failure will be eight times higher than the stress required to cause failure in simple uniaxial tension. It is indeed true that the strength of glass is eight times higher in simple compression than in simple tension.

## 7.4 BRITTLE FRACTURE IN CRYSTALLINE MATERIALS

The Griffith theory applies only to completely brittle materials. In crystalline materials which fracture in an apparently brittle fashion, plastic deformation usually occurs next to the fracture surface. Any theory of brittle fracture in such materials must, therefore, take into account not only the energy necessary to produce new crack surface energy but also that necessary to produce plastic deformation in the vicinity of the crack. Orowan has shown that when plastic deformation is concentrated in a region whose thickness is small compared to the length of a crack, the work of plastic deformation may be treated as a contribution to the effective surface energy of this crack. Recognizing this, Griffith's equation may be modified as:

$$\sigma = \left[\frac{2E(\gamma + p)}{\pi c}\right]^{1/2} \simeq \left(\frac{Ep}{c}\right)^{1/2} \tag{7.7}$$

where $p$ is the work of plastic deformation at the tip of the growing crack. Values for the plastic work term in low-carbon steel which have failed in a brittle manner range from $10^5$ to $10^6$ ergs/cm$^2$, approximately one thousand times greater than the true surface energy. When the modified Griffith equation is applied to those crystalline materials which show some degree of ductility, it is found that the critical crack length necessary for brittle fracture is of the order of a few millimeters instead of the few microns necessary in amorphous materials.

At low temperatures or high strain rates, brittle fracture may occur in those crystalline materials which exhibit rapid increases in their yield strengths with decreasing temperature and increasing strain rate, for example, certain BCC metals (iron, molybdenum and tungsten) and certain ionic salts (LiF, NaCl and KI). In these materials, in which some degree of plastic deformation precedes brittle fracture, the fracture usually results from pre-existing cracks, as in amorphous materials, and also from the formation of microcracks during slip or twinning. The stress concentration at the tip of a newly formed microcrack can be initially accommodated by plastic deformation. When the Griffith-Orowan criterion (Equation 7.7) is met, however, the crack can begin propagating spontaneously. The speed at which this spontaneous

*Table 7.1 Cleavage Planes in Several Materials**

| MATERIAL | CRYSTAL STRUCTURE | CLEAVAGE PLANE |
|---|---|---|
| MgO | rock salt | {100} |
| LiF | rock salt | {100} |
| NaCl | rock salt | {100} |
| NaBr | rock salt | {100} |
| KCl | rock salt | {100} |
| KBr | rock salt | {100} |
| KI | rock salt | {100} |
| PbS | rock salt | {100} |
| $CaF_2$ | fluorite | {111} |
| ZnS | zinc blende | {110} |
| InS | zinc blende | {110} |
| Ge | diamond | {111} |
| Si | diamond | {111} |
| C | diamond | {111} |
| $\alpha$–Fe | BCC | {100} |
| W | BCC | {100} |
| Be | HCP | {0001} |
| Zn | HCP | {0001} |
| C | graphite | {0001} |
| Te | selenium | $\{10\bar{1}0\}$ |

* After J. J. Gilman, "Cleavage, Ductility and Tenacity in Crystals" in *Fracture,* edited by B. L. Averbach et al. (See reference, p. 164.)

propagation occurs increases rapidly from zero to a limiting velocity which is about one-third the speed of longitudinal sound waves in the particular material. If the yield stress is strongly dependent on strain rate, the velocity of the crack increases to the extent that plastic deformation cannot accommodate the stress concentration at the head of the crack, and the crack spreads in a brittle manner.

In single crystals which exhibit brittle behavior, fracture occurs along definite crystallographic planes called cleavage planes, several of which are listed in Table 7.1. In polycrystalline materials, fracture can propagate along either cleavage planes or along grain boundaries. Transgranular cleavage fracture is more often observed in pure polycrystalline materials that fail in a brittle manner. Because the orientation of cleavage planes changes across grain boundaries, the propagation of fracture across such a boundary is difficult just as slip across a boundary is difficult. A crack spreads from one grain to another by nucleating a new crack in an adjoining grain; the two cracks then join by a tearing action which produces steps in the fracture surface near the grain boundary. In polycrystalline materials containing grain boundary segregates, fracture is usually intergranular. This mode of fracture has been observed in tungsten and molybdenum alloys containing oxygen, carbon, or nitrogen, copper containing bismuth or antimony, and iron containing phosphorus.

## 7.5 DUCTILE FRACTURE

Ductile fracture occurs after appreciable plastic deformation. In high-purity single crystals and polycrystals of very ductile materials, necking occurs, and localized plastic deformation continues until the sample has necked to a point or line. This type of fracture (rupture) is not observed in most polycrystalline materials. Ductile fracture of engineering materials in tension has three distinct stages: (1) the sample begins necking and cavities form in the necked region; (2) the cavities begin to coalesce into a crack in the center of the sample, and the crack proceeds outward toward the surface of the sample in a net direction perpendicular to the applied stress; and (3) the crack spreads to the surface of the sample in a direction 45° to the tensile axis. The result of this series of processes is the "cup and cone" fracture, shown in Figure 7.1*a*.

Once a tensile sample has begun to neck, both the stress and the deformation are concentrated in the necked region. The stress within this region is not a uniform tensile stress. It is, instead, a complex distribution of triaxial stress, as shown in Figure 7.4. The outer fibers are no longer sustaining a simple tensile stress. Ductile fracture begins at the center of the necked region where both the shear stress and the tensile stress are as high or higher than any other point in the sample. Cavities form, usually at inclusions (see Figure 7.5), and are elongated in the direction of the maximum shear stress. This elongation leads to further necking and the ultimate rupture of the materials which join and form

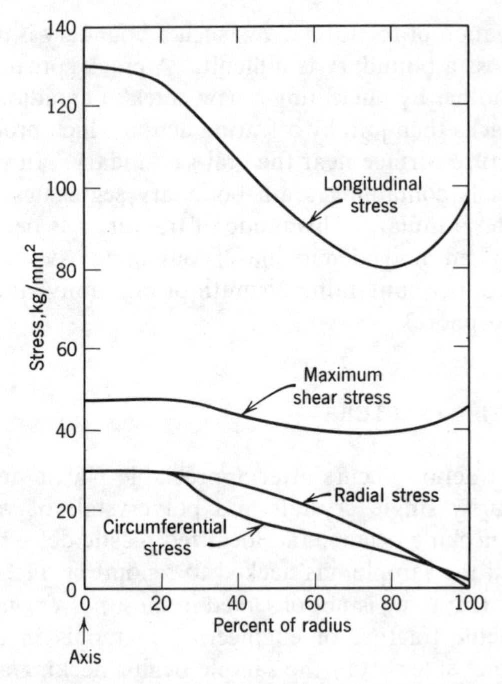


Figure 7.4 Stress distribution in the necked region of a sample of 0.25 per cent carbon steel pulled to fracture at room temperature (after E. R. Parker, H. R. Davis, and A. E. Flanigan, *ASTM,* 1946, p. 1159).

a central crack, as shown in Figure 7.5. The "cup" part of the fracture is produced by cracks traveling in a zigzag fashion at angles of 45° to the tensile axis. As a crack grows longer, it is necessarily carried away from the region of minimum cross-sectional area and highest stress, and thus after traveling for some distance at one 45° angle it reverses direction and goes back to the region of minimum cross-sectional area, again at a 45° angle. The "cone" part of the fracture is formed when the crack finally progresses close enough to the surface of the sample that the sample separates by shear fracture at an angle of 45° to the tensile axis. Although the "cone" portion of the fracture appears, macroscopically, to be smoother than the "cup" portion, electron microscopy shows that both are made by the same process of coalescence of cavities.

The ductility of an alloy, as measured by the reduction of area at fracture, is greatly affected by the volume of voids or poorly bonded inclusions existent in the alloy. Figure 7.7 illustrates how the ductility of copper in tension varies with the volume fraction of various included particles. Although other alloy systems have not yet been investigated as thoroughly as copper has, it is reasonable to expect similar behavior.

## 7.6 THEORIES OF CRACK INITIATION

The most effective area for crack initiation in alloys is in the vicinity of an included particle. When a metal undergoing deformation flows past an undeformable inclusion, dislocations pile up near the inclusion-metal interface. The tensile stresses associated with these dislocation pile-ups could lead to either shearing of the inclusion or creation of a void at the interface. Such voids form most readily at those inclusions which adhere most weakly to the

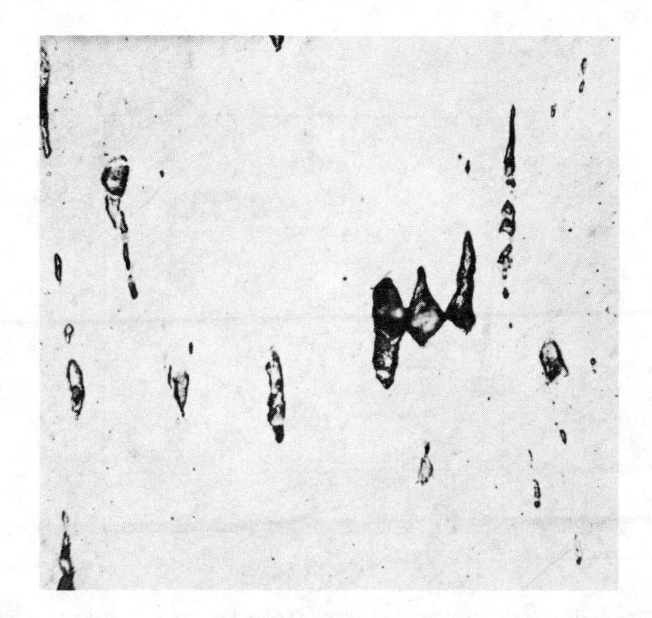

Figure 7.5 Cavities formed at oxide inclusions in copper (courtesy K. E. Puttick).

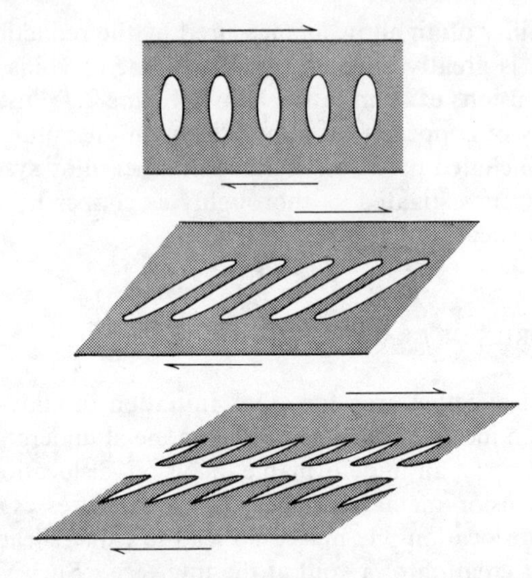

Figure 7.6 Schematic representation of production of "elongated dimples" on a shear fracture surface (courtesy W. A. Backofen).

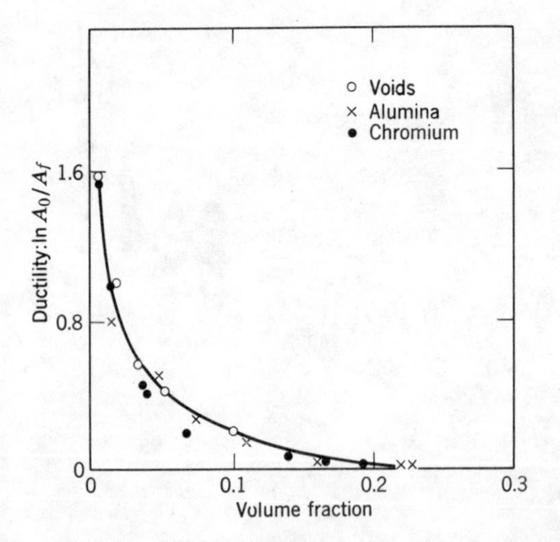


Figure 7.7 Ductility of copper as a function of volume of included voids, alumina particles, and chromium particles (data from Edelson and Baldwin, *Trans. ASM,* 55, 1962, p. 230).

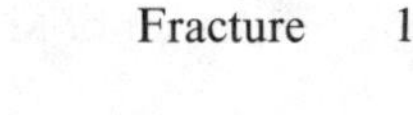

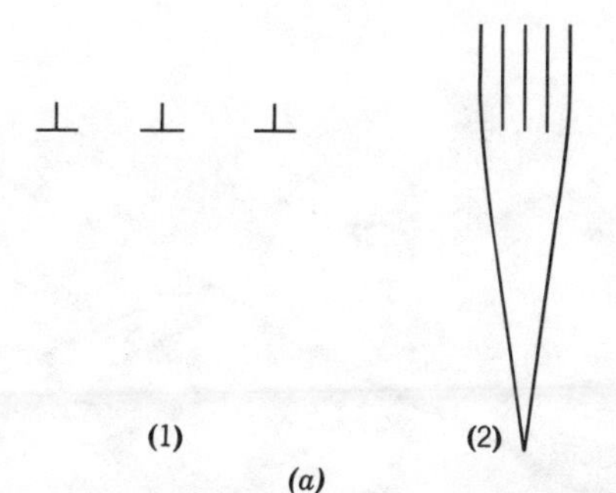


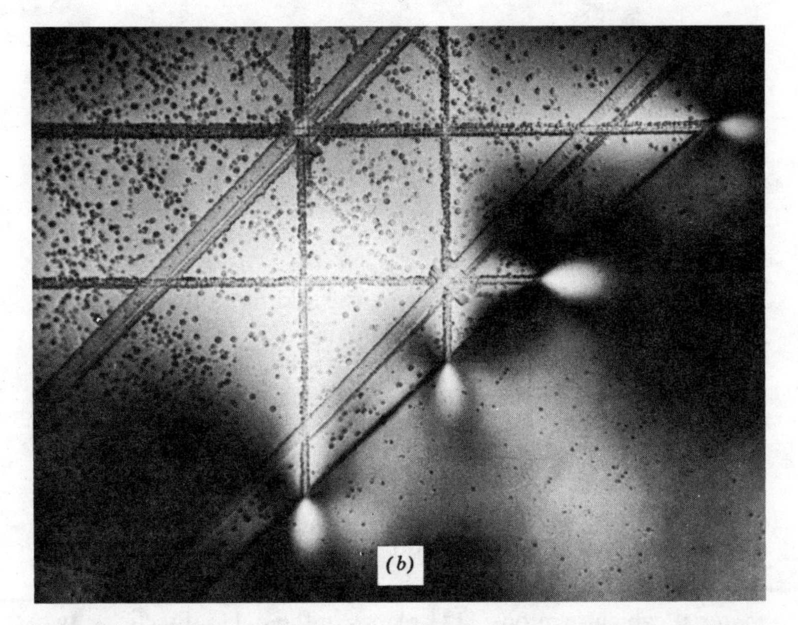


Figure 7.8 (*a*) Zener mechanism of crack formation by coalescence of dislocations piled up at a barrier to slip; (*b*) Stress concentrations produced at the grain boundary of an MgO bicrystal by blocked glide bands (courtesy T. L. Johnston).

matrix: for instance, voids form more easily at copper oxide inclusions in copper than at aluminum oxide inclusions in aluminum, for the adhesion between copper and copper oxide is much weaker than that between aluminum and aluminum oxide.

Zener first suggested that dislocation pile-ups could coalesce at any barrier to slip such as an inclusion or grain boundary, as shown in Figure 7.8*a*. The dislocations at the head of such a

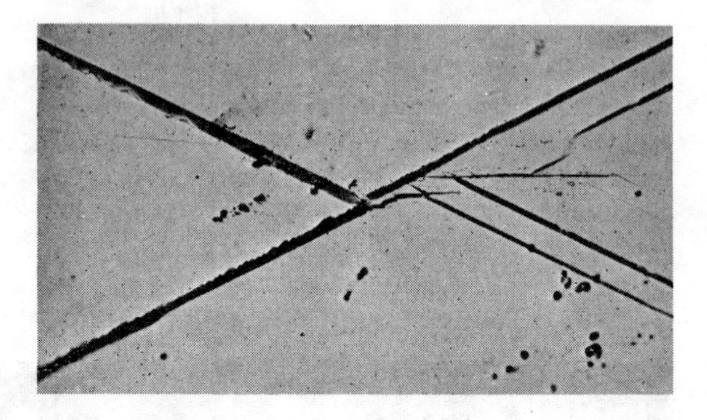

Figure 7.9 Cleavage crack nucleation at twin intersection in silicon-iron single crystals (from D. Hull, *Acta. Met.*, 8, 1960).

pile-up can become so closely spaced that a microcrack is nucleated, and this crack then grows by the addition of the remaining dislocations in the pile-up.

A large sessile dislocation may also serve in crack nucleation. Accordingly, the sessile dislocation, which is a barrier to further slip on either of the original slip systems, leads to pile-ups which then coalesce to form a microcrack having a $\{100\}$ cleavage plane. Although this is a possible and frequently discussed mechanism of crack nucleation, it is seldom observed.

Another method of microcrack formation concerns the successive intersections of deformation twins. In body-centered cubic material, which twin on $\{112\}$ planes in the $\langle 111 \rangle$ directions, twin intersection can lead to $\langle 100 \rangle$ microcracks when the tensile axis is in either the $\langle 011 \rangle$ or between the $\langle \bar{1}12 \rangle$ or $\langle \bar{1}11 \rangle$ directions. Continued growth of twins after they intersect lead to formation of a $\{001\}$ crack. A crack formed by this mechanism is shown in Figure 7.9. A similar mechanism occurs in high-temperature creep tests, where some of the over-all deformation is caused by sliding along grain boundaries. Here a crack forms in a grain boundary which is normal to the applied tensile stress and intersects two other grain boundaries along which some sliding has already occurred, as shown in Figure 7.10*a* and *b*.

Microcracks can also nucleate along low-angle tilt boundaries as shown in Figure 7.11. Because of the incompatibility of lattice

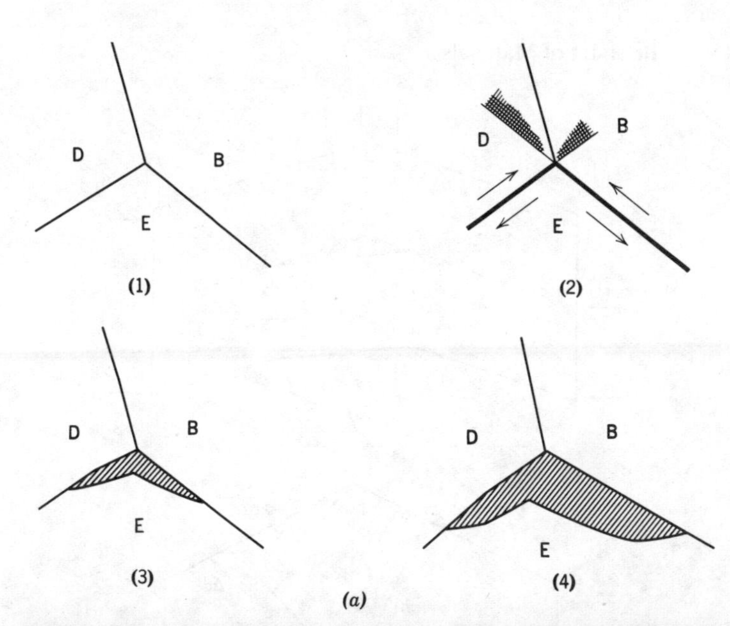


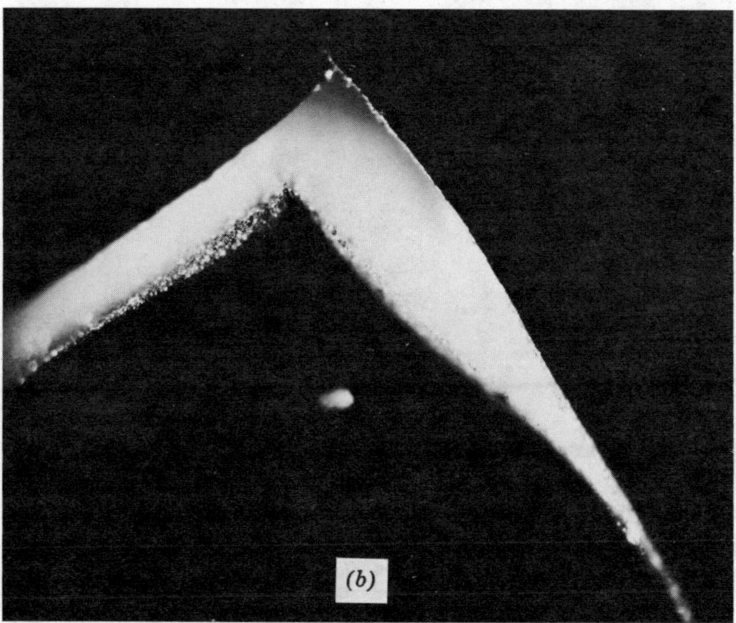


Figure 7.10 (*a*) Schematic model of intergranular cracking nucleated by grain boundary sliding. (1) before deformation; (2) sliding, especially in grain D; (3) initiation of intercrystalline crack at triple point; (4) growth of crack. The tensile axis is in the vertical direction. (*b*) Intergranular crack in an aluminum alloy creep specimen (from H. C. Chang and N. J. Grant, *Trans. AIME,* 206, 1956, p. 554).

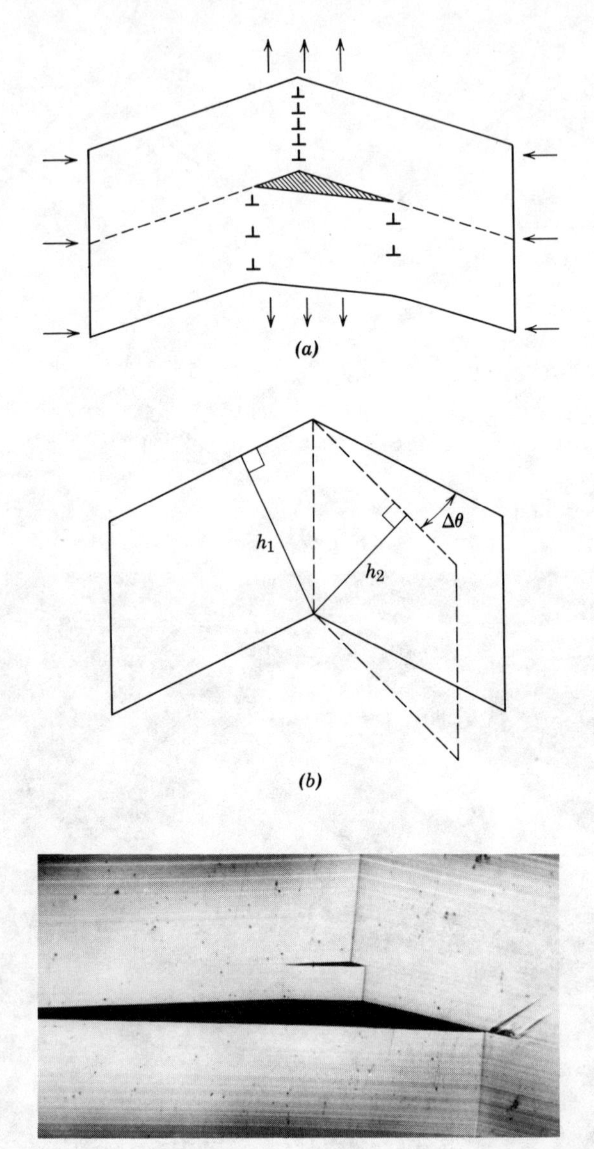


Figure 7.11 (*a*) Schematic mechanism of the formation of microcracks at tilt boundaries. (*b*) Formation of microcracks in zinc (from J. J. Gilman, *Trans. AIME,* 106, 1956, p. 1326). Specimen zinc crystal slowly deformed by compression roughly parallel to cracks while immersed in liquid $N_2$, 85×.

rotations, when slip occurs on a single plane, microcracks can be produced on that plane. This is a likely method of fracture initiation in HCP metals, where the cleavage and slip planes are the same.

## 7.7 DUCTILE TO BRITTLE TRANSITION

A transition from ductile to brittle fracture may be observed in BCC metals on decreasing the temperature, increasing the strain rate, or notching the material. The notched bar impact test described in Section 1.5 can be used to determine the temperature range over which the transition takes place. In this test, the determination of the transition temperature depends on (1) the transition in energy absorbed, (2) the transition in ductility, (3) the change in fracture appearance, and (4) the contraction at the root of the notch. The top curve of Figure 7.12 shows the transition based on energy absorption, the middle curve indicates a fracture transition, and the bottom curve a ductility transition. It is quite evident that the transition temperature is not a sharply defined temperature, and tests on the same lot of material exhibit appreciable scatter. In general, the sharper the notch, the higher the transition temperature. This is evident in Figure 7.12, where curves for the V-notch Charpy specimens are compared with keyhole charpy specimens. For steels and V-notched specimens the transition temperature is taken as that at which 10 or 15 ft-lb of energy are absorbed. The transition temperature is often taken to be that at which 50 percent fibrous (shear) fracture is obtained. When using a ductility criterion, the transition temperature is arbitrarily set for a 1 percent lateral contraction at the notch. In general, the fracture criterion for estimating transition temperature usually gives a higher transition temperature than that obtained using energy or ductility parameters. The scatter in transition temperature measurements is nevertheless great and depends to a large extent on testing conditions.

The importance of a notch in the impact test specimens is of significance in interpreting the origin of brittle failure. The notch provides a stress concentration and a constraint to plastic defor-

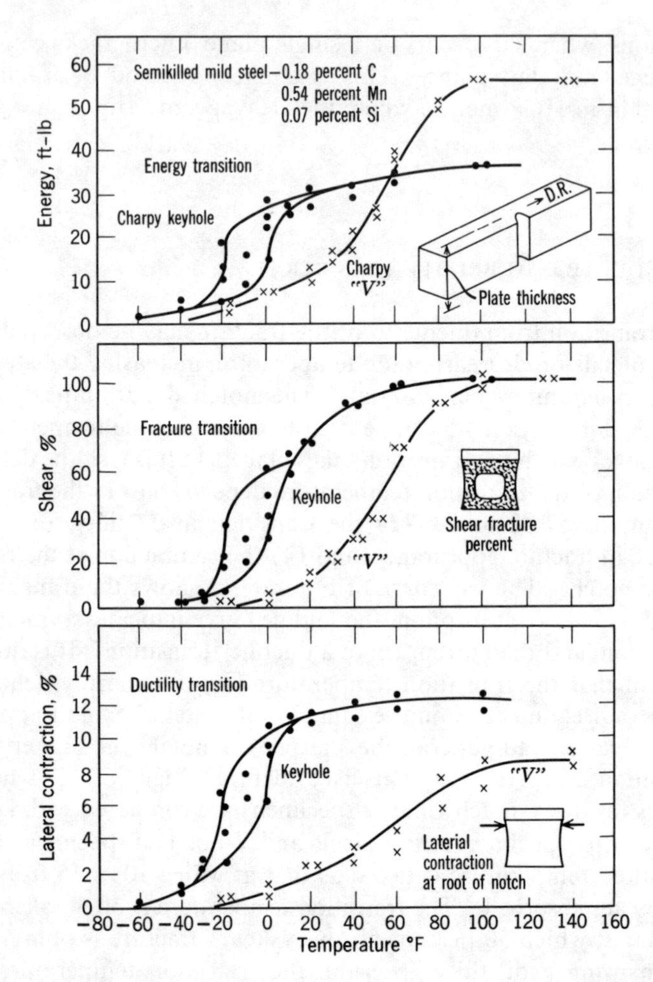


Figure 7.12 Transition-temperature curves based on energy absorbed, fracture appearance, and notch ductility (from W. S. Pellini, *ASTM Spec. Tech. Publ.*, 158, 1954, p. 222).

mation at its tip. When a load is applied, the notched region is in a state of triaxial tension. Slip or ductile deformation, which requires shear, will be suppressed by such a state of stress. Thus a notched specimen can support a much higher level of true tensile stress than an unnotched one. As a notched specimen is loaded high enough in simple tension, the material at the notch will try to yield. It would like to do so shrinking inward in the plane perpendicular to the applied tensile stress. The metal lying above and below the notch has not yielded and prevents yielding of the notch material. The latter is therefore subjected to three tensile stresses (triaxial). The first is the applied tensile stress, and the other two are induced horizontal tension stresses. If a mild steel tensile specimen were unnotched and subjected to the same applied true tensile stress as a notched one, it would yield at less than half the true tensile load sustained by the notched specimen. In the case of an ideally deep and sharp notch, the true tensile stress necessary to cause yielding in a notched specimen is three times as high as that for an unnotched specimen. The notch thus provides plastic constraint. The triaxial state of stress that it induces restrains plastic deformation and favors brittle fracture.

In materials whose yield stress increases sharply with increasing strain rate or decreasing temperature, for example, low carbon steel, the presence of notches and microcracks can induce brittleness. At intermediate temperatures, only after the crack tip reaches a rather high velocity does brittle failure occur. The high strain rate at the head of a rapidly propagating crack has then coupled with the plastic constraint in raising the yield stress above that necessary for cleavage (brittle fracture). Thus, if a very sharp notch is made in a steel plate and it is torn apart under tension, plastic deformation at the root of the notch will be found under all testing conditions. This plastic deformation takes the form of a small indentation and is of sufficient magnitude that in plates one-half inch thick that had atomically sharp notches, the resulting ripple can actually be felt on the surface of the sample.

## 7.8 FATIGUE FRACTURE

Ductile materials subjected to cyclic stresses much lower than their static fracture strength can fail by "fatigue." Observation indicates that a fatigue crack is induced which grows essentially by some form of plastic deformation. A characteristic fatigue fracture surface as shown in Figure 7.13 contains two distinct

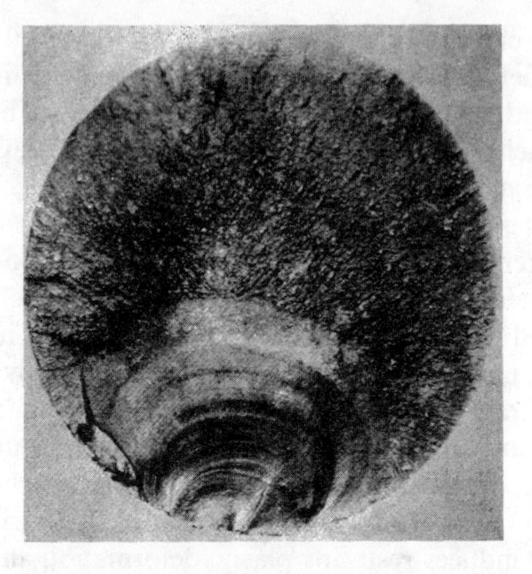

Figure 7.13 Typical fatigue fracture surface. Fracture originated at the point on the bottom edge of the sample from which the clamshell markings emanate (from B. Chalmers, *Physical Metallurgy,* John Wiley and Sons, 1959, p. 212).

zones: (1) a rather smooth region with concentric "clamshell markings" and (2) a region of granular fracture. The smooth region marks the slow crack propagation region. The clamshell or ripple markings indicate the crack stops before the final rapid propagation to failure portrayed by the granular region. The latter failure process takes up only the final 10 percent of the total life of the specimen.

The level of maximum stress before fatigue failure is called the *endurance limit.* It is roughly proportional to the UTS (ultimate tensile strength) of a metal and is defined as that cyclic stress below which failure does not occur for any number of cycles. The endurance limit of ferrous alloys is generally half the tensile strength. In nonferrous alloys it may be estimated to be as low as $\frac{1}{3}$ of the UTS. At such low stresses slip bands can often be seen on polished specimens, whose grains are favorably oriented, in the first 5 percent of their fatigue life. Near *stress raisers* such markings appear even earlier. At the most intense slip bands, surface irregularities gradually form. If they resemble depressions they are called *intrusions.* If they simulate eruptions they are called *extrusions.* A photomicrograph of this effect is shown in Figure 7.14. Fatigue cracks have been observed to form along both kinds of markings. Cracks of this kind tend to coalesce until one becomes large enough to initiate failure. If such a disturbed surface is removed by etching during an interrupted fatigue test, subsequent fatigue life is increased. Precipitation due to strain aging, since it arrests slip, can also improve the fatigue life of a metal. Alloys which strengthen in this manner can be incubated or "coaxed" at a higher amplitude of applied stress or at a higher temperature to hasten strain aging.

It is well known that if the surface of a specimen is softer than the interior, fatigue failures will occur earlier than if the reverse is true. Thus one of the ways to increase fatigue life is to surface-harden the material. Carburizing, nitriding, and shot peening all improve the fatigue strength of a material; plated coatings harm it.

No adequate all-encompassing theory of fatigue has as yet been found. Subtle changes in surface structure and topography with and without dislocation models have been considered but with limited success. One of the most puzzling questions, namely, why certain metals exhibit an *S-N* curve (stress versus number of cycles to failure) with a well-defined fatigue limit whereas others

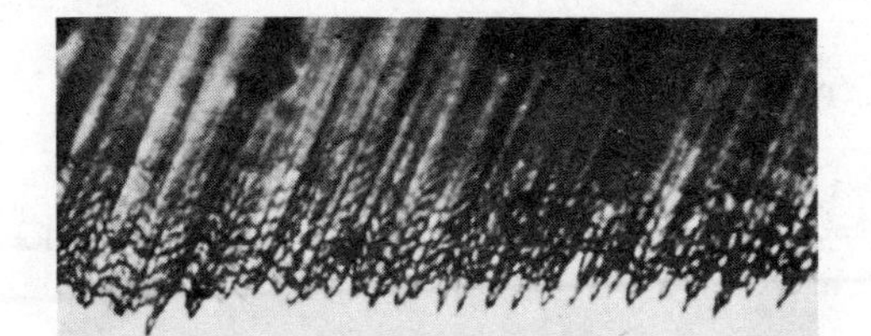

Figure 7.14 Formation of extrusions and intrusions in a sample of Cu-Al alloy tested in fatigue (courtesy D. H. Avery).

do not show any such limit, has long been the subject of study. It is believed at present that in general only materials which exhibit strain aging have a fatigue limit. However, an exception is heat-treated steel, which does not show strain aging in the tension test but which does indicate an endurance limit in fatigue tests.

## DEFINITIONS

*Brittle Fracture.* A mode of fracture characterized by the nucleation and rapid propagation of a crack with little accompanying plastic deformation. Brittle fracture surfaces in crystalline materials can be identified by their shiny, granular appearance.

*Ductile Fracture.* A mode of fracture characterized by slow crack propagation. Ductile fracture usually follows a zigzag path along planes on which a maximum resolved shear stress occurred. Ductile fracture surfaces usually have a dull, fibrous appearance.

*Cleavage Planes.* The characteristic crystallographic planes along which brittle fracture propagates. They are usually the planes having the lowest surface energy.

*Stress Concentration.* The magnification of the level of an applied stress in the region of a notch, void, or inclusion.

*Plastic Constraint.* The inhibition of plastic flow arising from the development of a triaxial stress distribution.

*Extrusions and Intrusions.* Eruptions and depressions on the surfaces of fatigue samples.

## BIBLIOGRAPHY

INTRODUCTORY READING:

Guy, A. G., *Physical Metallurgy for Engineers,* Addison-Wesley, Reading, Mass., 1962, Chapters 6 and 7.

SUPPLEMENTARY READING:

Cottrell, A. H., *Mechanical Properties of Matter,* John Wiley and Sons, New York, 1964.

Reed-Hill, R., *Physical Metallurgy Principles,* Van Nostrand, Princeton, N.J., 1964.

Smith, C. S., "Grains, Phases, and Interfaces," *Trans. A.I.M.E.,* **175** (1948), 15.

ADVANCED READING:

Averbach et al. (Eds.), *Conference on Fracture,* M.I.T. Press and John Wiley and Sons, New York and Cambridge, Mass., 1959.

Drucker, D. C. and J. J. Gilman (Eds.), *Conference on Fracture of Solids,* Interscience, New York, 1963.

## PROBLEMS

7.1 (a) The strength of graphite whiskers is up to $3.5 \times 10^6$ psi, that of quartz fiber up to $3.5 \times 10^6$ psi, and that of iron whiskers up to $1.9 \times 10^6$ psi. Piano wire has a strength of up to $0.35 \times 10^6$ and ausformed steel up to $0.45 \times 10^6$ psi, whereas hardwood along the grain has a strength of $0.015 \times 10^6$ psi. Calculate the ideal fracture strength, $\sigma_t$ of iron if

$$\sigma_t \simeq \sqrt{E\gamma/b}$$

where $E \simeq 2 \times 10^{12}$ dyn cm$^{-2}$, $\gamma \simeq 2000$ erg cm$^{-2}$, and $b \simeq 2.5 \times 10^{-8}$ cm (equilibrium spacing).

(b) Calculate the ideal fracture strength of NaCl if we take $\gamma$ for the cube plane to be 150 erg cm$^{-2}$, $E \simeq 5 \times 10^{11}$ dyn cm$^{-2}$, and $b \simeq 2.8 \times 10^{-8}$ cm.

7.2 (a) Indicate with actual experimental data the influence of surface finish on fatigue life of steel.

(b) Do the same for corrosive atmospheres.

7.3 (a) What is meant by "season cracking" of brass?

(b) How can it be ascertained and how avoided?

7.4 Write a short essay on stress corrosion.

(a) Distinguish between cleavage and shear fracture.

(b) Which predominates?

(c) When does a shear fracture indicate high ductility?

7.6 A steel whose grain size is ASTM No. 3, whose yield stress is 40,000 psi, fails in a brittle fashion by impact at 50,000 psi. Calculate the surface energy necessary to propagate a cleavage crack through the steel.

7.7 What is the stress required to spread a crack of length 0.1 mm in steel?

7.8 Consider a single diamond scratch whose depth is 0.002 in. in the transverse direction at midlength of a 20 in. $\times$ 10 in. $\times$ .20 in. glass sheet. Assuming the modulus of the glass is $8 \times 10^6$ psi, calculate the bending moment required to break the glass into two parts.

7.9 (a) Explain the practical use of the term "toughness" in materials testing. Does this apply to fracture?

(b) In what ways can size affect fracture toughness?

7.10 What evidence can you present that ductile cup-and-cone fracture is probably caused by piled-up groups of defects at grain boundaries at the end of coarse slip lines?

7.11 Explain why recrystallized commercial tungsten and molybdenum are both brittle at room temperatures, whereas recrystallized iron is ductile.

7.12 (a) With the aid of a bar of rectangular cross section explain how residual stress can arise.

(b) How can residual stress be removed from a metal?

(c) How does its presence affect fracture?

7.13 Write a short essay on radiation damage explaining its nature. Emphasize its effect on yield stress and on fracture.

7.14 Explain the cracking of a metal due to heat shock and thermal fatigue.

7.15 Explain why the strength of nearly all glass articles can be increased by etching off a thin surface layer with hydrofluoric acid.

7.16 A glass cube one cc in volume has two sharp surface cracks (one $\mu$ in length). Calculate its strength when tested in tension.

$$E = 10 \times 10^6 \text{ psi} = 7 \times 10^{11} \text{ dynes/cm}^2$$
$$\gamma = 300 \text{ ergs/cm}^2$$
$$\nu = 0.25 \text{ (Poisson's ratio)}$$

(a) If this piece is tested in compression with perfect lubrication, so that there is no frictional constraint between the platens and the test piece, what will be its compressive strength?

(b) If there is frictional constraint, will the compressive strength be higher or lower? Why?

7.17 Repeat Problem 7.16 for the case of a metal which fails in a brittle manner but shows intense plastic deformation in the vicinity of the cracks.

$$E = 20 \times 10^{11} \text{ dynes/cm}^2$$
$$p = 10^6 \text{ ergs/cm}^2$$
$$\nu = 0.28.$$

7.18 Discuss the validity of the statement: "In brittle materials the larger the specimen, the lower the fracture strength."

7.19 Some materials which fail with no plastic extension when tested in tension are observed to have some ductility when tested in compression or in pure shear. Why might this be so?

7.20 Discuss why the presence of notches does not always cause lower fracture strengths in all materials.

7.21 Why would the Zener method of crack nucleation not seem to be a probable mechanism in FCC metals having very high stacking-fault energies? Would the Hull mechanism of fracture initiated by twin intersections seem any more probable?

7.22 Why are materials whose yield stresses are highly strain-rate dependent more susceptible to brittle fracture than those materials whose yield stresses do not show marked strain-rate dependence?

7.23 On the basis of the data shown in Figure 7.12 and the discussion in Section 7.7, draw a schematic curve of toughness as a function of temperature for mild steel. (Toughness $= \int_0^{\epsilon_f} \sigma \, d\epsilon$.)

7.24 Why does a "keyhole-notched" impact sample exhibit a lower ductile-to-brittle transition than a "V-notched" sample?

7.25 Why is a shiny, smooth area near the point of crack initiation a characteristic feature of fatigue fracture surfaces?

7.26 In FCC materials, discuss how the magnitude of the stacking fault energy can affect fatigue life.

7.27 Why does the appearance of a "brittle" fracture surface not necessarily mean that the sample has not exhibited ductility prior to fracture?

7.28 In a short paragraph, discuss why the observed fracture strengths of both brittle and ductile materials are, with few exceptions, usually much lower than their theoretical strengths.

7.29 Consider a composite material consisting of a hard second phase dispersed in a soft ductile matrix. Discuss how the matrix phase can act to inhibit or aid crack propagation in the specimen.

7.30 It is found that electroplated chromium coatings on steel, though hard, do not improve the fatigue strength of steel unless the steel is initially shot-peened. Explain.

CHAPTER EIGHT

# *Strengthening Mechanisms*

Numerous theories for the hardening and strengthening of solids have been advanced. New materials have been developed with and without such knowledge. Some of the theories and the materials have been introduced in Volumes I and II and strain-hardening in particular in previous sections of this volume. The present chapter extends the description of some of the mechanisms involved and the processes employed.

## 8.1 INTRODUCTION

There is little doubt that *structure* plays an important role in the mechanical behavior of solids. It depends first of all on *chemical composition* and then on *mechanical* as well as *thermal processing.* In the latter are included casting, sintering, hot-working, and heat treatments of all kinds. Such processing steps influence mechanical properties by their effect on grain size, concentration gradients, inclusions, voids, metastable phases, dispersed phases, and lattice imperfections of different kinds. As pointed out previously, strain-hardening due to the accumulation of dislocations during plastic deformation is one of the major modes of strengthening a metal. Solid solution alloying is another. In regard to the latter, *interstitial* and *ordered* solid *solutions* as well as *clustering* before precipitation from a supersaturated solid solution are worth distinguishing. Different *dispersion* hardened materials as well as *transformation* and *diffusion* hardened materials also deserve description. Much of metallurgical progress has been registered in terms of how each of these various mechanisms has been employed in developing ever stronger alloys.

## 8.2 COLD-WORKING AND ANNEALING

In rolling, wire drawing, and other cold-work processes about 90 percent of the expended energy is dissipated as heat. The remainder is stored in the lattice as an increase in internal energy. It amounts to about 0.01 to 1.0 cal/g of metal and is greater the higher the melting point and alloy content of the metal. The stored energy increases with strain up to a saturation value. It also increases with decreasing temperature of working. Some of the stored energy may be attributed to vacancy formation and some to twin and stacking fault energy. In general, however, the major part of the stored energy is due to the generation and interaction of dislocations. The density of the latter is increased from $10^6$ to $10^{12}$ per square centimeter as we go from a fully annealed to a severely cold-worked metal.

Hardness and yield strength, in general, increase with cold work. In these processes the grains tend to become elongated and take on a *preferred* crystallographic *orientation* ("texture"). Severe cold deformation of a metal results in a slight decrease in density. The electrical conductivity decreases and the thermal expansion increases. Of greater importance, however, is the general increase in chemical reactivity of the cold-worked state which leads to a higher rate of corrosion. Severely cold-worked brass exhibits the phenomenon of "season cracking" a type of *stress corrosion cracking* when immersed in mercurous chloride or ammoniacal solutions.

The harmful effects of cold work, as has been mentioned previously, can be removed by heat treatment. In heat treatment of a severely cold-worked material it is customary to distinguish three temperature regions, namely, recovery, recrystallization, and grain growth. In recovery, heat treatment changes in grain structure are not observable. Electrical conductivity becomes similar to that of annealed metal, and X-rays definitely indicate that lattice strain is reduced. The latter fact accounts for the widespread use of recovery heat treatments.

The driving force for recovery and recrystallization is the stored energy of cold work. As the temperature of heat treatment is raised through the recovery range, strains are relieved and polygonization of the structure occurs. This is due to dislocations grouping themselves into the lower energy configurations of low angle grain boundaries. By dislocation climb a polygonal network of low angle grain boundaries results.

At a somewhat higher temperature than that required for recovery, a replacement of the cold-work structure by new strain-free grains starts to take place. This marks the onset of recrystallization. Recrystallization is readily detected mechanically because hardness and strength drop sharply and ductility increases. As expected, the density of dislocations decreases appreciably during recrystallization. Any further increase in temperature of heat treatment beyond the recrystallization range rapidly enlarges the grain size.

From a practical standpoint the recrystallization temperature is taken to be the temperature at which a totally new grain structure will appear in one hour. Strictly speaking, it is *not* a *fixed* temperature but *a range* whose lower temperature depends on such variables as initial grain size, composition, amount of previous cold work, temperature cycle, and time.

## 8.3 SOLUTE-HARDENING

A common way to increase the hardness and yield strength of a metal and particularly its strain-hardening rate is by *solid-solution alloying*. Figure 8.1 indicates the effect of various solutes on the yield strength (at 1 percent proof stress) of copper. The effectiveness of the impurity addition depends on the *size difference* and the percentage present. If the atom is larger than the solvent atoms, compressive strain fields are set up, and if it is smaller, tensile fields. The presence of either type impedes dislocation motion. Recent theoretical studies have extended our understanding of the influence of small amounts of solute atoms. These take both the change of lattice parameter ($\varepsilon_b$) and the shear modulus of the immediate surroundings ($\varepsilon_G$), into account. In copper the increase of yield stress per unit concentration for substitutional elements plotted against an elastic modulus parameter ($\varepsilon_s$) gives a straight line from Zn to As to Sn, as shown in Figure 8.2; the latter atom has the greatest effect.

In alloys where the increase in yield stress per unit concentration of solute is of the order of one-tenth of the shear modulus, hardening is considered to be "slow." Where it is a few times

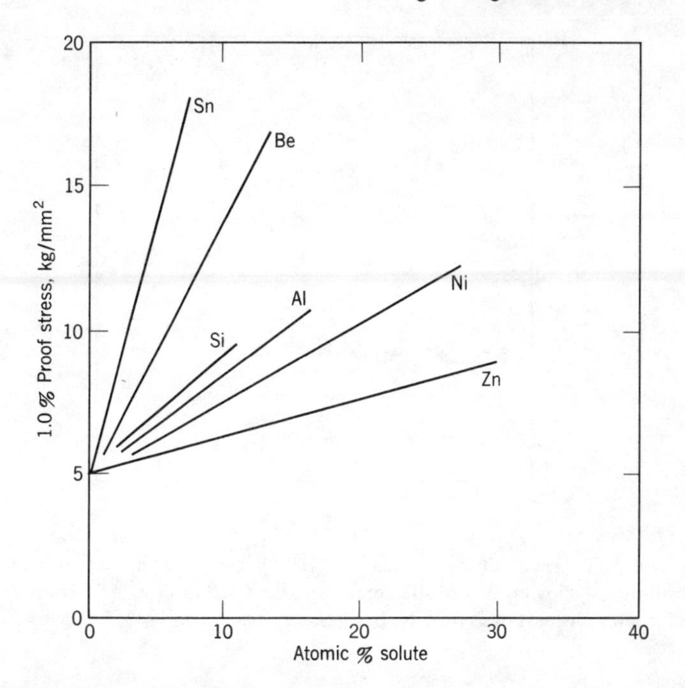


Figure 8.1 Effect of soluble alloy elements on the 1.0 percent proof stress in copper polycrystals at room temperature (from R. S. French and W. R. Hibbard, *Trans. AIME,* 1950, p. 53).

greater than the shear modulus, it is considered rapid. Impurities which distort the lattice in tetragonal fashion cause rapid hardening. This may be accomplished by a divalent ion vacancy pair, as in the NaCl lattice, by an interstitial atom in a BCC metal, or by a vacancy disc. The rapid hardening in these cases is due to a stronger interaction between tetragonal distortion and screw dislocations.

The above view of solute hardening also applies to dilute solid solutions. It is then necessary to consider an interaction between the faulted area of extended dislocations and solute atoms responsible for hardening. For more concentrated solid solutions local ordering could impede slip. Short range or local order exists in an alloy if the number of like neighbor atoms is different from that expected statistically. The first dislocation that passes a slip

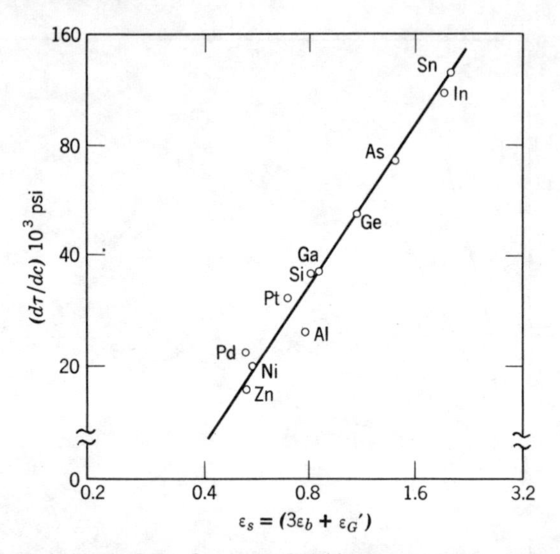


Figure 8.2 Relation between increase of yield stress per unit concentration for substitutional alloying elements in copper and the total misfit parameter appropriate for screw dislocations (from R. L. Fleischer, *Acta. Met.*, 9, 1961, p. 996).

plane experiences a retarding force. Only after extremely large deformation does short-range order cease to play a role. Large deformations probably destroy the local order.

Long-range order is different from short-range order. In the former, atoms of one kind occupy preferred lattice sites in the host-lattice, thereby producing a superlattice. Thus correlation occurs over many lattice sites. If heat treatment produces perfect ordering, there is little increase in hardness. If, however, a disordered phase is first produced by quenching from high temperature, and some ordering is induced by heat treatment at some lower predetermined temperature, a sizable increase in strength is possible. In 18 carat gold order-hardening can be accomplished by precipitation of the ordered AuCu phase. Similar but more extensive hardening occurs in 50 atomic percent Fe-Rh alloys. On quenching from the disordered state a 50 atomic percent Co-Pt alloy a Vickers hardness of 175 is obtained but on aging for 1 hour at 700°C its hardness increases to 325. Vicalloy (35 percent Fe, 52 percent Co, 13 percent V) is a permanent magnet alloy that hardens by ordering.

Although the yield stress of a fully ordered alloy is seldom higher than that of the disordered structure, it work-hardens more rapidly than the disordered alloy, as shown in Figure 8.3. This effect is attributed to the presence of antiphase boundaries. These are formed between ordered regions within a grain and resemble the grain boundaries in a polycrystalline metal. Dislocations are retarded at such antiphase boundaries. Since the motion of dislocations also increases the amount of antiphase boundary, the retarding force increases with increasing deformation.

## 8.4 PRECIPITATION-HARDENING

In certain alloy systems, precipitation-hardening can be far more effective than hardening by reduction of grain size, cold work, or solid-solution hardening. Alloys susceptible to precipitation-

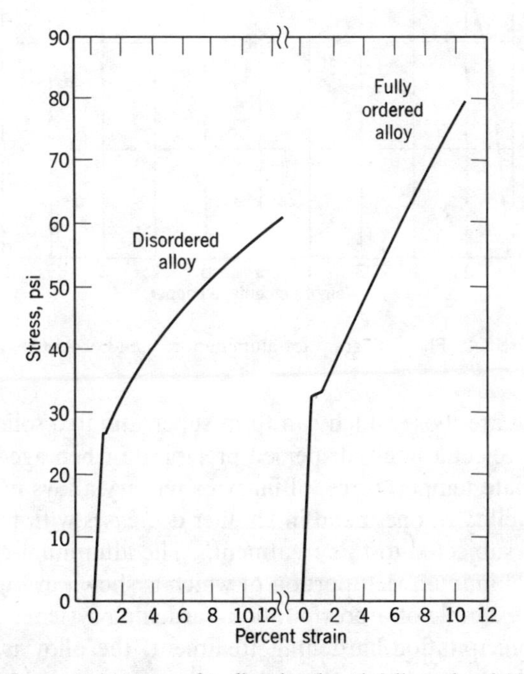


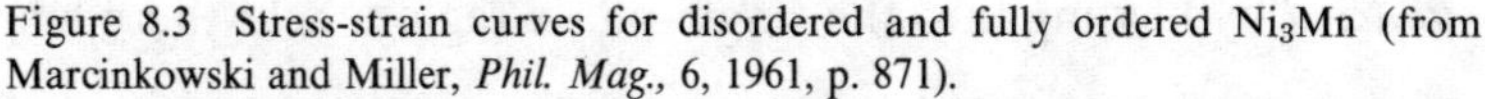

Figure 8.3 Stress-strain curves for disordered and fully ordered $Ni_3Mn$ (from Marcinkowski and Miller, *Phil. Mag.*, 6, 1961, p. 871).

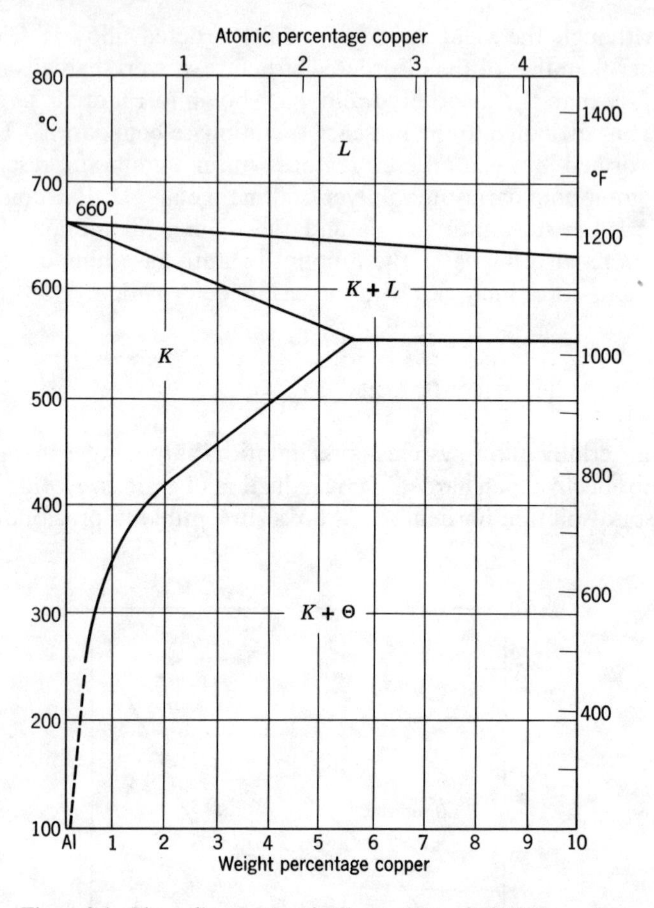


Figure 8.4 Phase diagram for aluminum-rich end of Al-Cu system.

hardening are those which can form supersaturated solid solutions and then reject a finely dispersed precipitate when aged at low or intermediate temperatures. Binary or ternary alloys in which the solid solubility of one metal in another decreases with temperature are often subjected to this treatment. The aluminum-copper system, the aluminum-rich portion of which is shown in Figure 8.4, is a typical example of a precipitation-hardening system.

In a precipitation-hardening treatment, the alloy is first *solutionized* by heating into the single-phase region, held there long enough to dissolve all existing soluble precipitate particles, and is then rapidly quenched into the two-phase region. The rapidity of the quench prevents the formation of equilibrium precipitates and thus produces a supersaturated solid solution. On aging at or above room temperature, fine-scale transition structures, as small as 100 Å, form. The amount of hardening produced by aging quenched aluminum-copper alloys of varying copper content is shown in Figure 8.5. The changes in hardness indicated in this figure result from the formation of three transition structures, illustrated schematically in Figure 8.6. These structures, denoted GP-1, GP-2 and $\theta'$, all form before the equilibrium phase $CuAl_2$. Their precipitation within the base aluminum leads to local distortions and strain fields which impair dislocation mobility.

The first two transition structures are called *Guinier-Preston* (*GP*) zones after the men who first studied their formation by X-ray diffraction. The *GP-1* zones are clusters of copper atoms about 100 Å long and two or three atoms thick, segregated on (100) planes in the aluminum matrix, and are completely *coherent* with the matrix (that is, there is a one-to-one alignment between the two kinds of atoms). Because there is a difference between the two atomic radii, complete coherency requires that elastic

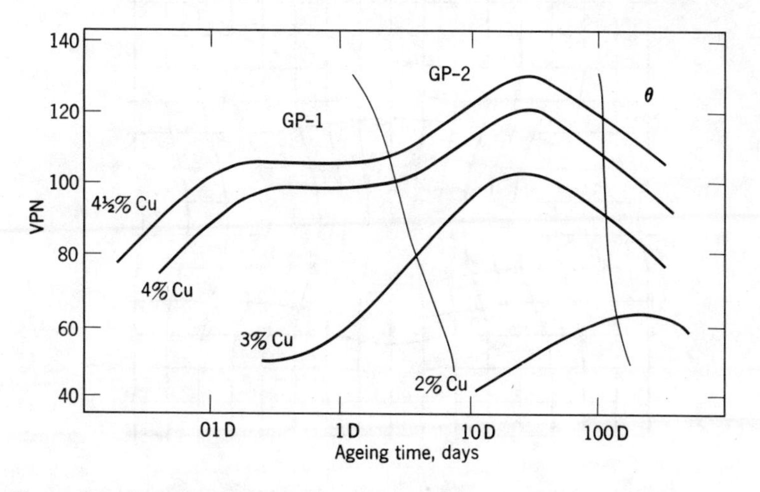


Figure 8.5 Variation of hardness with time for various Al-Cu alloys aged at 130°C (from Hardy and Heal, *Progress in Metal Physics,* 5, 1954, p. 195.)

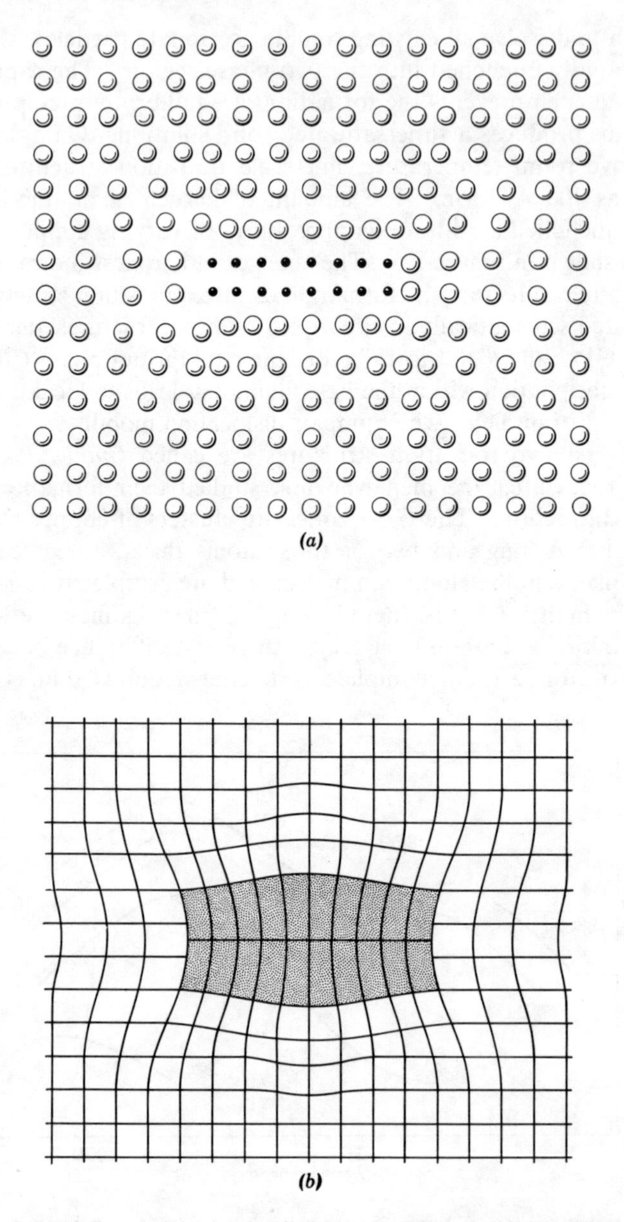


Figure 8.6 Al-Cu precipitation-hardening: (*a*) GP-1 zones; (*b*) GP-2 zones; (*c*) $\theta'$ structure; (*d*) $\theta$ structure.

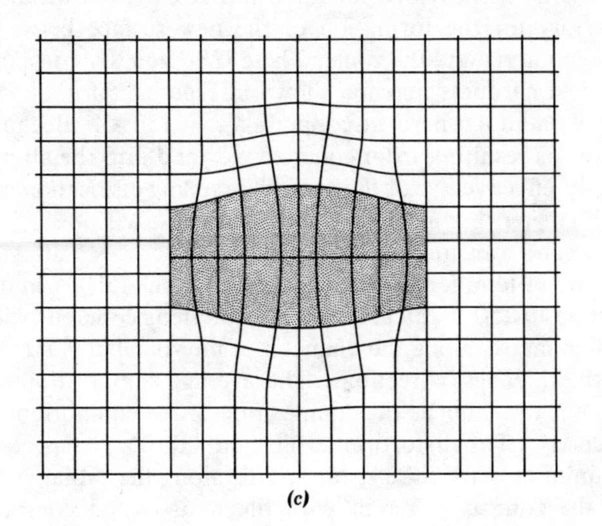

*(c)*

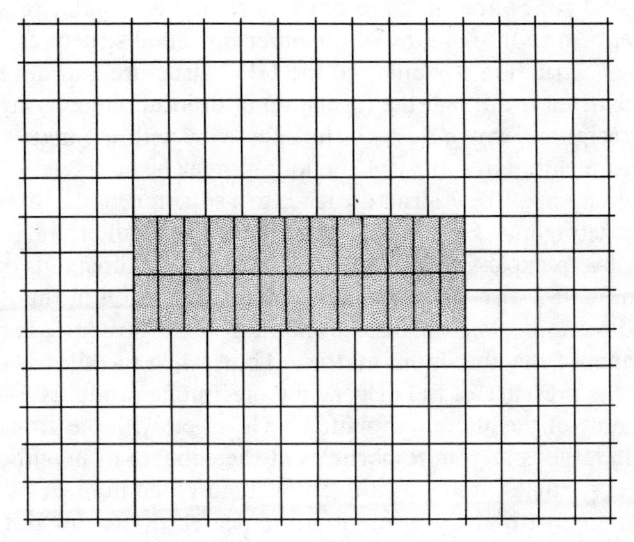

*(d)*

strain energy be provided for them to nucleate, but little energy is necessary for the formation of the new surface between the aluminum matrix and the zone. These *GP-1* zones are responsible for the first hardness maxima shown in Figure 8.5 for alloys containing 4.0 and 4.5 percent copper aged at 130°C. Because the elastic fields resulting from coherency extend into the aluminum lattice, the effective size of the zone in impeding dislocation motion is much greater than its actual physical size.

The second transition structure, the GP-2 zone, is a disc-shaped precipitate of tetragonal crystal structure about 1500 Å in diameter and up to 150 Å thick. This precipitate is coherent with the aluminum matrix along the diameter of the zone but is incoherent along the thickness direction. The average composition of this structure is the same as the composition of the equilibrium structure, $CuAl_2$. The distortions which the GP-2 zones produce in the aluminum matrix are shear strains along the radial surfaces, where the zone is coherent with the matrix, and compressive strains along the thickness of the zone, where it is incoherent with the matrix. Optimum hardness is attained with the formation of the GP-2 zones, for they are both more numerous and produce greater distortion than any of the other transition structures.

The $\theta'$ structure is similar to the GP-2 structure and develops directly from it through the formation of dislocation loops around the precipitate, which destroys its coherency with the matrix and therefore eliminates most of the long-range elastic forces. Thus the formation of the $\theta'$ structure leads to a softening of the material.

The tetragonal $CuAl_2$ ($\theta$) precipitate, the equilibrium phase, forms by an expansion of the $\theta'$ cell in the radial directions of the original GP-2 disc-shaped zones, and a contraction in the thickness direction. The intrinsic strength of the $\theta'$ structure is more than that of the aluminum matrix. Thus, when the alloy is overaged, the presence of the dispersed $\theta$ precipitate produces a slight hardening of the aluminum matrix. However, with the growth of certain larger precipitate particles at the expense of neighboring smaller particles, upon continued overaging, the number of sites interfering with dislocation motion decreases to the possible extreme that the hardness of a completely overaged alloy can be less than the hardness of the supersaturated solid solution which existed prior to the aging treatment.

While the Al-Cu system is taken to be a typical precipitation-hardening system it is well to remember that other systems may differ from it in detail. The general sequence: supersaturated solid solution → transition structure → equilibrium phase is nevertheless the same in all systems and therefore provides a lowering of free energy at all stages of precipitation.

## 8.5 DIFFUSION-HARDENING

Strengthening by a *fine dispersion* of obstacles can also be achieved by allowing selected gases to react with and diffuse into solids. This is the case when steel which has been first heat treated to have a tempered martensite structure is exposed to an ammonia atmosphere at 500 to 600°C for times of from 12 to 36 hours. If the steel contains small additions of such alloying elements as Al, Cr, and V, a fine dispersion of *nitrides* is formed which markedly increases surface hardness and wear resistance. Prolonged use of such materials at temperatures above 500°C causes first of all a reduction in hardness due to loss in coherency and gradual coalescence of fine particles.

*Internal oxidation* of cold-rolled silver sheet containing small amounts of Al or Mg in solid solution leads to surface hardening as well as strengthening. Due to the solubility of oxygen in silver, it diffuses into the sheet and forms a fine precipitate of MgO or $Al_2O_3$ at 700 to 800°C. This reduces the electrical restivity and raises the hardness and yield strength more than threefold. Such material is at present used for electrical contacts.

## 8.6 MARTENSITIC TRANSFORMATIONS

The *martensitic* transformation is a *diffusionless,* displacive reaction and occurs in systems in which a diffusion-controlled *invariant transformation* can be suppressed by rapid cooling. As already mentioned in Volumes I and II, it occurs by a process of lattice shearing; the martensitic phase is formed from the retained high-temperature phase at temperatures lower than the equilibrium invariant transformation temperature. These reactions represent the tendency of a system to assume a crystalline structure which more closely resembles the equilibrium structure than does the retained, high-temperature phase.

As viewed in the microscope, martensite appears as *lenticular* plates which divide and subdivide the grains of the parent phase, always touching but never crossing one another (see Figure 8.7). The lenticular shape minimizes the elastic distortion in the matrix caused by the martensite platelet.

Another characteristic feature of the martensite reaction is the great speed at which the platelets grow: at about one-third the velocity of sound. This high velocity of growth indicates that the activation energy for platelet growth is very low; as a consequence the activation energy for nucleation determines the amount of martensite formed under a given set of conditions. Martensite generally starts to form on cooling the parent phase below a critical temperature, $T_s$, and is finished at a low temperature $T_f$. The effect of cooling below $T_s$ is to increase the driving force for the martensitic transformation, thus enhancing formation of nucleation sites of a size which can now grow. Under mechanical deformation, martensite can sometimes be formed at temperatures higher than the $T_s$.

Martensite platelets attain their shape by two successive shear displacements contained in boundaries coherent with the parent phase (see Figure 8.8). The first displacement is a homogeneous shear throughout the plate which occurs parallel to a specific plane in the parent phase known as the habit plane. The second displacement, the lesser of the two, can take place by one of two mechanisms: slip, as in Fe-C martensite, or twinning, as in Fe-Ni martensite. Martensite transformations occur in a large number of alloy systems, including Fe-C, Fe-Ni, Fe-Ni-C, Fe-Mn, Cu-Zn, Au-Cd, and even in pure metals like Li, Zr, and Co.

## 8.7 IRON-CARBON MARTENSITE

Martensite in ferrous alloys containing carbon is unique for two reasons: (1) the martensite structure is distorted into an asymmetric (body-centered tetragonal) structure by interstitial carbon atoms, and (2) the as-formed martensite is extremely hard and

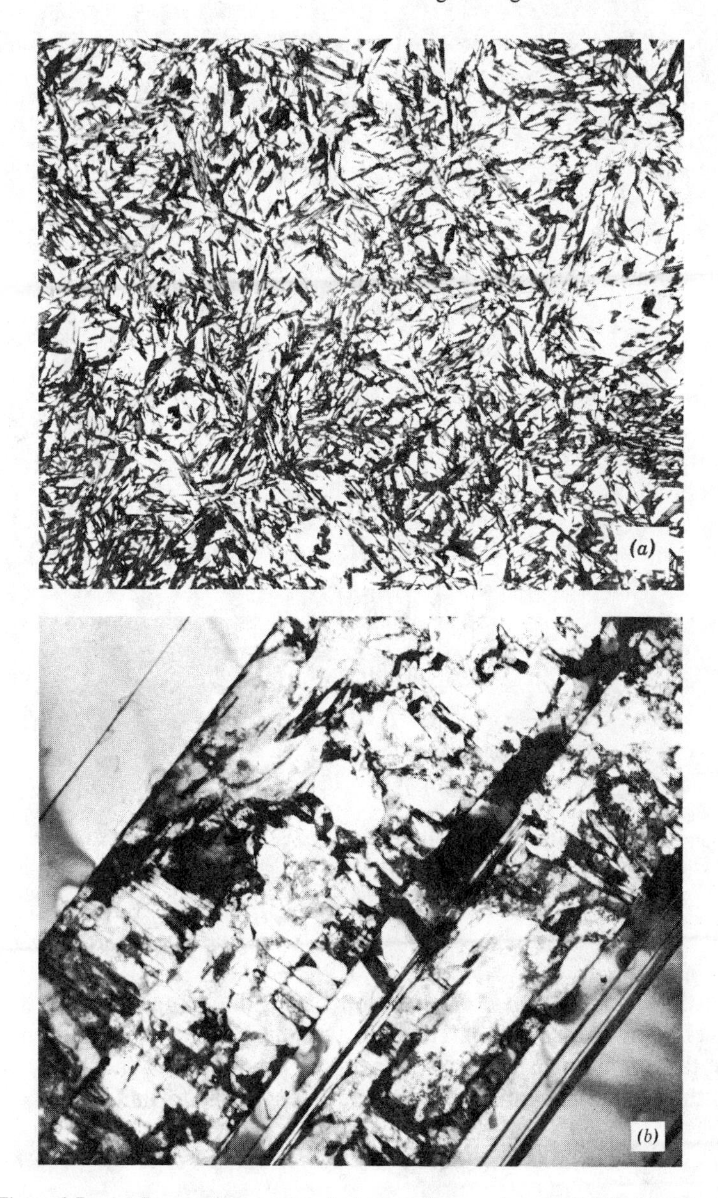


Figure 8.7 (*a*) Iron-carbon martensite in a medium-carbon steel, 370× (courtesy N. S. Pitea, International Nickel Company). (*b*) Martensite formed on quenching an alloy of Fe, 16% Cr, 12% Ni to −196°C, 7,100× (courtesy J. F. Breedis).

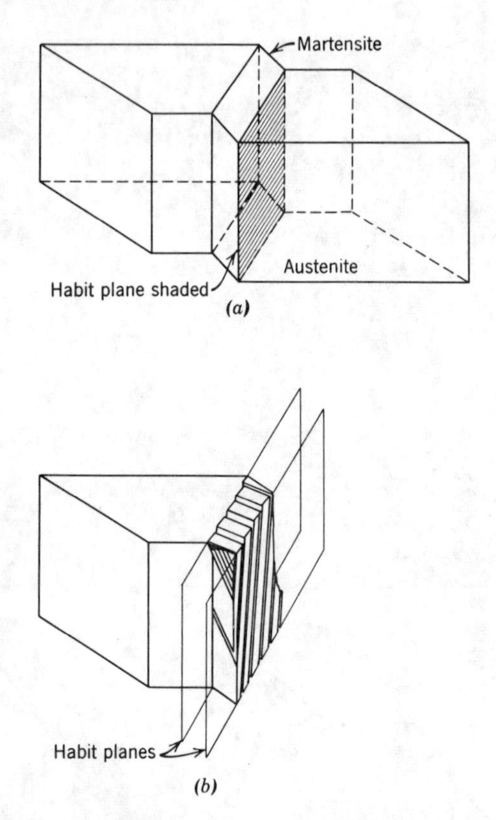


Figure 8.8 Schematic representation of two displacements by which a martensite platelet is formed: (*a*) primary displacement; (*b*) secondary displacement accommodated by fine-scale twinning. Secondary displacement may also be accommodated by fine-scale slip.

brittle. Both of these features depend on the composition of the steel being transformed. The tetragonality of the Fe-C martensite lattice, as measured by the ratio of the height of the unit cell to the base of the unit cell ($c/a$), is given by the formula:

$$\frac{c}{a} = 1.000\ (\pm 0.005) + 0.045 \times (\text{wt percent C}) \qquad (8.1)$$

Thus the degree of tetragonality is directly proportional to the amount of carbon in the austenite from which the martensite is formed. As the carbon content approaches zero, the $c/a$ ratio approaches unity. In this case, the resultant structure would be the body-centered cubic structure of ferrite. Hence Fe-C martensites may be considered to be supersaturated solid solutions of carbon in ferrite which have a tetragonal crystal structure.

The hardness of Fe-C martensite increases, although not linearly, with carbon content, as shown in Figure 8.9. The assymetric distortions of the martensite lattice by interstitial carbon atoms are similar to those produced by *GP* zones in age-hardenable alloys, and seem to be the most important cause of hardening.

When considering the strength of martensite, it is necessary to distinguish as-formed from tempered martensite. In the as-quenched martensitic state, the alloy is both too hard and too brittle for practical application. By tempering, ductility can be increased and hardness and strength decreased, making the alloy tougher and hence more useful. Tempering moves the martensitic structure toward quasi-equilibrium so that ferrite and cementite appear. Tempering of eutectoid composition (0.8 percent C) martensite generally occurs in four stages. During the first stage, at temperatures ranging from room temperature to 200°C, martensite having the carbon content of the alloy decomposes into a mixture of low-carbon martensite (0.3 percent C) and $\epsilon$ carbide,

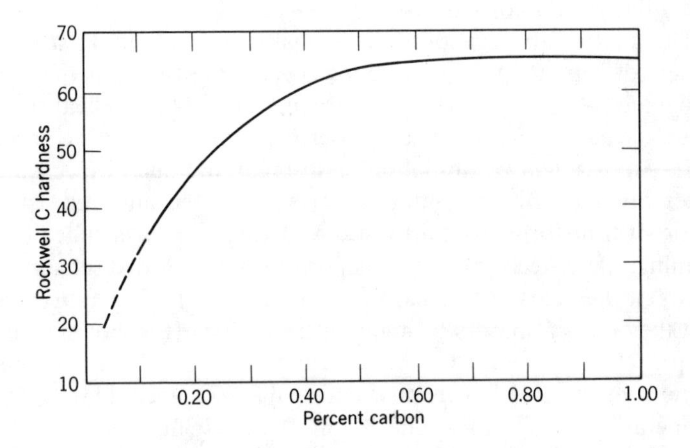


Figure 8.9 Maximum hardness versus carbon content for any completely hardened steel.

a transition precipitate whose composition ranges from $Fe_2C$ to $Fe_3C$. Because $\epsilon$ carbide is both a distorted structure and coherent with the low-carbon martensite matrix, the first stage of tempering produces a slight amount of hardening. In the second stage of tempering, in the temperature range 200 to 300°C, any retained austenite decomposes into bainite, a fine mixture of ferrite and cementite. This stage of tempering causes a slight degree of softening. During the third stage, occurring between 260 and 360°C, the low-carbon martensite and $\epsilon$ carbide decompose into ferrite and cementite, producing a marked softening of the alloy. The fourth stage, which occurs at high temperatures (up to the eutectoid temperature, 723°C), causes spheroidization of the carbide particles and the subsequent growth of large carbide particles at the expense of smaller ones. Although the total volume of cementite does not change during spheroidization, the reduction in the number of carbide particles produces softening in a manner similar to that observed in the overaging of precipitation-hardenable alloys.

In general, tempering produces minor strengthening and then softening, except in ferrous alloys containing such strong carbide formers as chromium, tungsten, and vanadium. Here martensites strengthen markedly on tempering through a process known as *secondary hardening,* which is probably due to the precipitation of finely dispersed alloy carbides.

The martensitic reaction is the basis for hardening a wide variety of commercial ferrous alloys. Typical of these is the steel designated *AISI 4340,* which is hardened by austenitic heat treatment, quenching and tempering after machining and before grinding. An excellent combination of strength and ductility is sometimes obtained by deforming a hot steel in the austenitic state before it transforms to martensite or bainite. This is called ausforming. The steels used for this purpose have a broad separation between the pearlite and bainite regions of their TTT diagrams, and they must also possess high hardenability. It is then possible to quench such steels to an intermediate temperature where austenite is metastable but readily deformable. Quenching from this temperature results in a fine, highly strained martensite. This martensite is then tempered. Tensile strengths some 35 percent

*Table 8.1 Composition and Mechanical Properties of Four High-Strength Iron Alloys*

| | AISI 4340 | VASCOMAX 300 | 17-4PH | H-11 |
|---|---|---|---|---|
| C | 0.40 | 0.03 max | 0.07 max | 0.40 |
| Si | 0.30 | 0.10 max | 1.00 max | 1.00 |
| Mn | 0.75 | 0.10 max | 1.00 | ... |
| Ni | 1.80 | 18.50 | 4.00 | ... |
| Cr | 0.80 | 0.05 | 16.50 | 5.00 |
| Mo | 0.25 | 4.80 | ... | 1.30 |
| Co | ... | 7.50 | ... | ... |
| Al | Trace | 0.10 | ... | ... |
| Ti | ... | 0.40 | ... | ... |
| B | ... | 0.003 | ... | ... |
| Zr | ... | 0.02 | ... | ... |
| Cb and Ta | ... | ... | 0.35 | ... |
| Cu | ... | ... | 4.00 | ... |
| V | ... | ... | ... | ... |
| Hardness as quenched, $R_c$ | 54–56 | 28–32 | 35 | 63 |
| Aging or tempering Temperature, °F | 400 | 900 | 900 | 1000 |
| *Properties after Aging or Tempering* | | | | |
| Hardness, $R_c$ | 52–54 | 50–54 | 44 | 62 |
| 0.2 percent offset yield stress, psi | 240,000 | 280,000 | 180,000 | 360,000 |
| Ultimate tensile strength, psi | 280,000 | 290,000 | 200,000 | 400,000 |
| Percent elongation | 12 | 12 | 14 | 8 |
| Percent reduction in area | 30 | 55 | 50 | 40 |
| Room temperature charpy impact energy, ft-lb | 13 | 17 | 25 | ... |
| Endurance limit, psi | 100,000 | 123,000 | 80,000 | 160,000 |

higher than those produced by conventional heat treatments of the same steel have been achieved by ausforming.

In iron nickel alloys of high enough nickel content, such as Vascomax, it is possible by quenching to obtain a relatively soft low-carbon martensite which can be worked and machined. The structure may then be precipitation-hardened to high tensile strengths without distortion. This process is called maraging. Table 8.1 gives the composition of a maraging steel. Another precipitation-hardening alloy called stainless steel 17-4PH is also listed in this table along with the ausforming alloy and AISI 4340.

## DEFINITIONS

*Cold-Working.* Any form of mechanical deformation processing carried out on a metal below its recrystallization temperature.

*Hot-Working.* Any form of mechanical deformation processing carried out on a metal or alloy above its recrystallization temperature.

*Work-Hardening.* Hardening and strengthening of a metal or alloy caused by plastic deformation.

*Age-Hardening.* The strengthening of an alloy resulting from the precipitation of a finely dispersed second phase from a supersaturated solid solution.

*Overaging.* A phenomenon resulting from the coarsening of the precipitates in an age-hardenable alloy. At this stage, the precipitates usually attain the equilibrium structure, and with further coarsening, the strength of the alloy decreases markedly.

*Dispersion Strengthening.* A means of strengthening a metal by creating a fine dispersion of insoluble particles within the metal.

*Martensite Transformations.* Diffusionless crystallographic transformations which occur by displacive reactions in metals and alloys.

*Tempering.* For steels, the heat treatment by which iron-carbon martensites are toughened. With the precipitation of carbides and the concurrent lowering of the martensite carbon content, the strength of the alloy decreases but its ductility increases.

## BIBLIOGRAPHY

INTRODUCTORY READING:

Brick, R. M. and A. Phillips, *Structure and Properties of Alloys,* McGraw-Hill Book Co., New York, 1949, Chapters 6, 9, 10, 11, 12.

Guy, A. G., *Elements of Physical Metallurgy,* Addison-Wesley, Reading, Mass., 1959, Chapter 6.

Rogers, B. A., *The Nature of Metals,* American Society of Metals, Cleveland, 1951, all chapters.
Smallman, R. E., *Modern Physical Metallurgy,* Butterworth, London, 1962, all chapters.
Van Vlack, L. H., *Elements of Materials Science,* 2nd ed., Addison, Wesley, Reading, Mass., 1964, Chapters 1, 4, 6, 7, 8, 9, 10, 11.

SUPPLEMENTARY READING:

*A.S.M. Handbook,* Cleveland, 1961.
Averbach, B. L. and M. Cohen, *Trans. A.S.M.* **41** (1949), 1024.
Cottrell, A. H., *Mechanical Properties of Matter,* John Wiley and Sons, New York, 1964.
Dieter, G. E., *Mechanical Metallurgy,* McGraw-Hill Book Co., New York, 1961.
Fisher, J. C., Synthetic Microstructure, *Trans. A.S.M.* **55** (1962), 916.
Guard, R. W., "Mechanics of Fine Particle Strengthening," in *Strengthening Mechanisms in Solids,* A.S.M., Cleveland, 1962.
Hollomon, J. H. and L. D. Jaffe, *Ferrous Metallurgical Design,* John Wiley and Sons, New York, 1947.
McLean, D., *Mechanical Properties of Metals,* John Wiley and Sons, New York, 1962.
Read-Hill, R. E., *Physical Metallurgy Principles,* Van Nostrand, Princeton, 1964.
Roberts, C. S., B. L. Averbach, and M. Cohen, *Trans. A.S.M.* **45** (1953), 576.
Seal, A. K. and R. W. K. Honeycombe, *Journal of the Iron and Steel Institute* (London) **188** (1958), 9.
Smith, C. S., "Grains, Phases, and Interfaces," *Trans. A.I.M.E.* **175** (1948), 15.

## PROBLEMS

8.1 Single crystals are very much weaker than they theoretically should be, because dislocations can operate to produce slip at low values of resolved shear stress. In what way can the presence of grain boundaries in polycrystals lead to higher yield strengths than those of single crystals?

8.2 High purity lead can be worked by rolling almost indefinitely at room temperature without any noticeable hardening.

(a) Why might this be so?

(b) Would this also be the case at the boiling point of liquid nitrogen (77°K)?

8.3 What feature of a binary phase diagram is essential in determining if any of the alloys in the system are capable of precipitation-hardening?

8.4 Draw a schematic diagram of temperature versus time for properly hardening a precipitation-hardenable alloy.

8.5 Why is it preferable to precipitation-harden alloys such as Al-4.5 percent Cu by first quenching to a low temperature and then reheating to the aging temperature rather than by quenching directly to the aging temperature?

8.6 Given a cast slab of a 5 percent (Cu) Al-Cu alloy;

(a) Sketch its as-cast microstructure.

(b) State at what temperature you would anneal it to get rid of coring.

(c) How else might you process it to get a homogenized structure?

(d) Indicate with hardness versus time curves how you would select the optimum temperature and time of aging.

8.7 Why are aluminum alloy rivets refrigerated after solutionizing?

8.8 Discuss a possible use for an age-hardenable aluminum alloy which ages but does not overage (in a finite period of time) at room temperature.

8.9 What feature about an alloy or a pure metal causes it to undergo a martensitic transformation?

8.10 Why is the martensitic reaction in iron-carbon alloys unique among martensitic reactions?

8.11 Compare the structures and properties of austenite, ferrite, cementite, pearlite, and martensite in plain carbon steels.

8.12 Compare the effects on mechanical properties of rapidly quenching a eutectoid carbon steel and an age-hardenable alloy. How does the hardness of each respond to subsequent heat treatment?

8.13 Discuss the similarities between the tempering of iron-carbon martensite and the precipitation-hardening of an Al-Cu alloy.

8.14 (a) What is the composition of Hadfield Mn Steel?

(b) How is it heat treated to get all austenite?

(c) How can it be given a hard martensitic surface?

8.15 By what mechanism does 18:8 stainless steel harden during working?

8.16 Discuss in a brief essay the subject—maraging steel. Use the journals *Trans of A.S.M.* and *Metal Progress.*

8.17 Consider a laminated sheet composed of two different materials A and B. The volume fraction of material A is V. If the composite is stressed in tension parallel to the lamellae, show that the modulus of the composite (assuming equal strains in A and B) is given by $E = E_A V + (1 - V)E_B$.

CHAPTER NINE

# Ceramics and Other Inorganic Nonmetallics

The mechanical behavior of nonmetallics, like that of metals, depends in large measure on microstructure and macrostructure, which in turn depends on chemical composition, raw materials, and mode of fabrication. Nonmetallics may be formed from the melt or they may be fabricated by sintering or by cementing powder particles. Although nonmetallics are generally weak in tension, their strength in compression is often appreciable. They are also used on account of their resistance to creep at high temperatures, resistance to oxidation and corrosion, and their thermal and electric insulating capacity. In this chapter, the deformability, fabrication, strength, and thermal characteristics of inorganic nonmetallics are briefly considered.

## 9.1 INTRODUCTION

Nonmetallics are usually characterized by ionic, covalent, and intermediate bonding. They may exist as crystals, glasses, or gels (a gel is a colloidal suspension: a fine-scale mixture of solids, liquids, and frequently gases in which the solid portions exhibit distorted crystallinity). Such materials as silica ($SiO_2$) may be found in either of these three forms. As discussed in Volume I, an inorganic compound tends to be noncrystalline if (1) each anion is linked to not more than two cations, (2) no more than four anions surround a cation, (3) the anion polyhedra share corners, but not edges or faces, and (4) the material has many constituents. Unless a material capable of forming a glass is cooled extremely slowly from the liquid phase, a glassy structure having only *short-range order* will result.

For the most part, nonmetallic inorganic materials are brittle; however, under carefully controlled experimental procedures, certain crystalline forms of these materials may be quite ductile. If single crystals of sodium chloride are tested in tension after being stored in air for a prolonged period, they are brittle; when tested in water, they exhibit appreciable ductility. If allowed to remain in air for a short time after immersion, they become brittle again. Early interpretation attributed this effort to the removal of the surface cracks which induce brittleness by the dissolution of the surface. More recent work has suggested that variations in the moisture content of the atmosphere cause precipitates to form on the surface of the crystals by a process of local dissolution and subsequent reprecipitation. These precipitates are observed to be points at which fracture originates. Although ductility is possible and has been observed in some ionic crystals, most fabricated polycrystalline ceramics behave in a brittle fashion at low temperatures.

## 9.2 PLASTICITY OF SINGLE CRYSTALS

In ionic, nonmetallic single crystals plasticity is limited by geometric and electrostatic considerations. In sodium chloride structures (NaCl and MgO), as shown in Figure 9.1, slip generally occurs on the $\{110\}$ plane in the $\langle 110 \rangle$ direction. In this direction the amount of slip necessary to restore the structure is least. At high temperatures slip occurs in such structures in the plane $\{100\}$ and the direction $\langle 110 \rangle$.

If slip is confined to $\{110\}$ $\langle 110 \rangle$ systems there are only two independent slip systems. This is because the strain which results from the operation of any one slip system is identical to the strain which results from the operation of a perpendicular slip system and is also equivalent to the sum of the strains resulting from the operation of two slip systems which are not perpendicular to themselves or to the original slip system. This number of independent slip systems precludes polycrystalline ductility. Five independent systems are necessary for ductility in randomly ori-

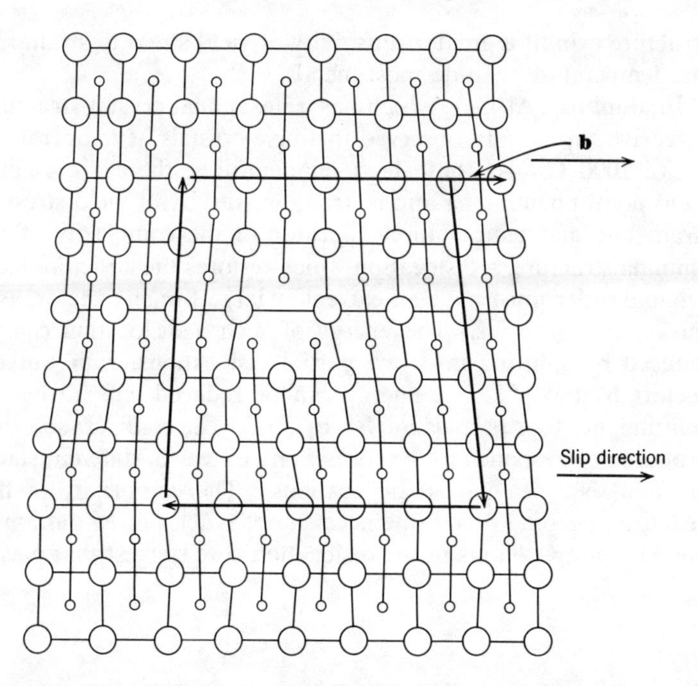


Figure 9.1 Edge dislocation in sodium chloride structures, showing slip direction, Burgers circuit, and Burgers vector, **b**.

ented, polycrystalline materials. In polycrystalline AgCl, where slip occurs readily on {110} and {100} planes, there is an adequate number of independent slip systems, and a degree of ductility comparable to that of ductile metals is observed.

In highly ionic materials, it has been found that grown-in dislocations are quite effectively locked by the presence of small amounts of impurities (a few parts per million). As a result, although the grown-in dislocation content might be quite high, the dislocations accounting for most of the plastic deformation observed are nucleated after the crystal has been grown. This gives rise to very pronounced yield point effects, indicating that the stress necessary for the production of movable dislocations is greater than that required for their subsequent motion. Aside from the effect of impurity locking, most ionic materials with the NaCl structure exhibit a greater sensitivity of yield stress to strain rate and temperature than do most metals.

In alumina ($Al_2O_3$), which has a hexagonal crystal structure, extensive slip is only observed in single crystals at temperatures above 1000° C. At these high temperatures, there are definite yield point phenomena and a strong sensitivity of yield stress to strain rate and temperature. Because of the complexity of the alumina structure, a dislocation which restores the structure to its original order must have an extremely large Burgers vector, as is shown in Figure 9.2. The energy of such a dislocation can be reduced by splitting into two partial dislocations, with Burgers vectors $\mathbf{b}'$ and $\mathbf{b}''$. The energy can be reduced still further by splitting into four partials $\mathbf{b}_1'$, $\mathbf{b}_2'$, $\mathbf{b}_1''$, $\mathbf{b}_2''$. The configuration thus produced corresponds to the formation of areas of different stacking fault between partial dislocations. The complexity of this dislocation geometry and the necessity of synchronized motion of the $Al^{+3}$ and $O^{-2}$ ions in the dislocation core suggest that plastic

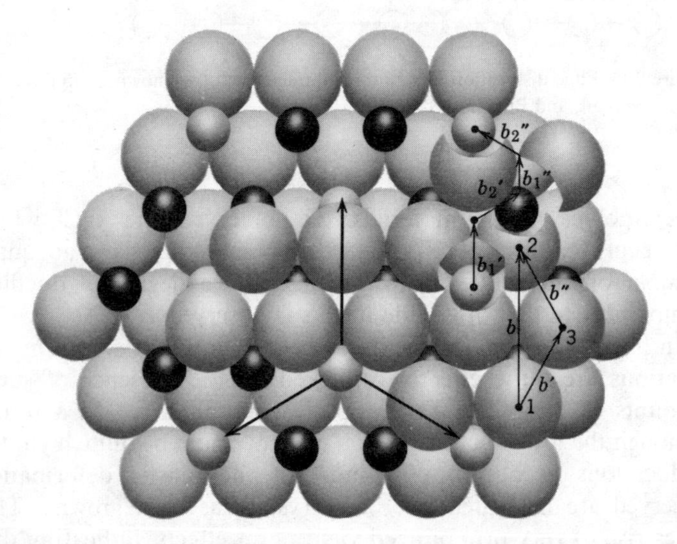


Figure 9.2 Structure of $Al_2O_3$, showing two layers of large oxygen ions with hexagonal array of $Al^{+3}$ and vacant octahedral interstices. Slip directions and Burgers vector for basal-plane slip are indicated (from M. L. Kronberg, *Acta Met.*, 5, 1957, p. 507).

deformation in alumina can occur only when the strain rate is low and the temperature is high enough so that the ions are relatively mobile.

Most ceramic materials have an insufficient number of independent slip systems for general deformation and may have structures which are so complex that dislocation motion is difficult. However, the controlling factor limiting ductility is the ease of fracture. If fracture can be suppressed, crystalline ceramics may deform plastically as higher stresses move dislocations on slip systems other than the preferred ones. These additional slip systems can provide the degrees of freedom needed for general deformation.

## 9.3 VISCOUS BEHAVIOR OF GLASSES

The mode of deformation in ceramic materials is highly dependent on structure. As we discussed in the previous section, the nature of any possible deformation in crystalline ceramics results from plastic slip processes. In glassy ceramics, the lack of any *long-range order* makes dislocation motion impossible. However, these materials can deform by viscous processes under proper conditions of stress and temperature.

A fluid flows viscously, that is, it cannot support an applied shear stress statically, and thus it deforms continuously. The resulting shear strain $\gamma$ is then a function of both shear stress $\tau$ and time $t$:

$$\gamma = f(\tau, t) \tag{9.1}$$

For an ideal, Newtonian fluid,

$$\tau = \eta \frac{\partial \gamma}{\partial t} = \eta \dot{\gamma} \qquad (\eta = \text{constant}) \tag{9.2}$$

when $\eta$ is the coefficient of viscosity, usually measured in poises.

Many noncrystalline solids obey Equation 9.2 to a reasonably good approximation; examples are inorganic glasses and organic glasses (noncrystalline polymers) at relatively high temperatures. In addition, lower temperature behavior of these materials can usually be rationalized in terms of some combination of pure elastic deformation and *viscous* (not necessarily ideal, or Newtonian) flow.

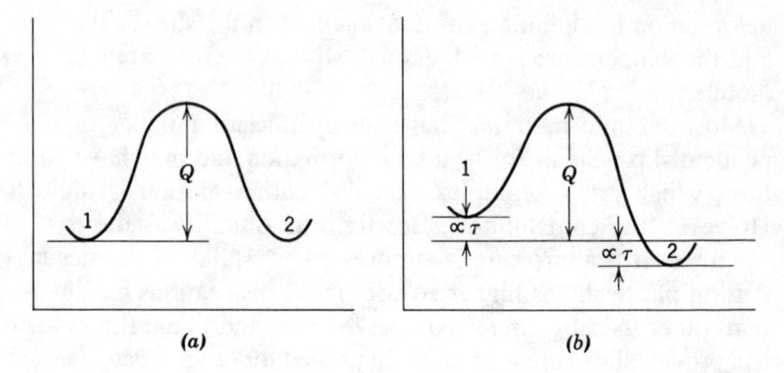


Figure 9.3 A hypothesized activation energy barrier for viscous flow (*a*) before and (*b*) after applying a stress $\tau$. The energy $\alpha\tau$ is a very small fraction of the energy $Q$.

If we assume that the temperature is high enough so that deformation is completely viscous, it is possible to derive Equation 9.2 as a special case of a kinetic rate process controlled by an activation energy barrier. Consider a model in which each of the units which shear past one another (atoms, ions, or molecules, as the case may be) under an applied stress has two metastable equilibrium positions which do not differ significantly in energy until a stress is applied (Figure 9.3).

If the shear strain rate, $\dot{\gamma}$, of the material is assumed to be proportional to the rate at which units migrate over the activation energy barrier:

$$\dot{\gamma} = C[N_1 e^{-(Q-\alpha\tau)/kT} - N_2 e^{-(Q+\alpha\tau)/kT}] \tag{9.3}$$

where $N_1$ is the number of units in position 1, $N_2$ is the number of units in position 2, $Q$ is the activation energy per atom, $k$ is Boltzmann's constant, and $T$ is the absolute temperature.

$$\dot{\gamma} = Ce^{-Q/kT}[N_1 e^{\alpha\tau/kT} - N_2 e^{-\alpha\tau/kT}] \tag{9.4}$$

Since $\alpha\tau$ is very small compared to $Q$ (this is equivalent to the earlier statement, that the process is primarily thermally activated), the first two terms of the exponential expansion

$$e^x = 1 + \frac{x}{1!} + \cdots \tag{9.5}$$

can be used for the quantities in brackets in Equation 9.4:

$$\dot{\gamma} = Ce^{-Q/kT}\left[N_1\left(1 + \frac{\alpha\tau}{kT}\right) - N_2\left(1 - \frac{\alpha\tau}{kT}\right)\right] \tag{9.6}$$

At equilibrium, $N_1 \simeq N_2$ because $(Q - \alpha\tau) \simeq (Q + \alpha\tau)$, so to an approximation

$$\dot{\gamma} = Ce^{-Q/kT}\left[\frac{2N\alpha\tau}{kT}\right] = \left[\frac{K}{T}e^{-Q/kT}\right]\tau \tag{9.7}$$

Combining with Equation 9.2,

$$\frac{1}{\eta} = \frac{Ke^{-Q/kT}}{\tau} \quad \text{where} \quad K = \frac{2N\alpha}{k} \tag{9.8}$$

$$\eta = T \cdot K'e^{Q/kT} \quad \text{where} \quad K' = \frac{1}{K} \tag{9.9}$$

Since the exponential term dominates the right-hand side of Equation 9.9, the *coefficient of viscosity,* $\eta$, should increase rapidly with decreasing temperature. An example of this type of behavior is shown in Figure 9.4 for a sodium silicate glass (74.5 percent $SiO_2$). The plot of log $\eta$ versus $1/T$ in Figure 9.4*b* deviates from a straight line in the direction predicted by the temperature factor in the pre-exponential part of Equation 9.9, but over small temperature intervals the curve is approximately linear.

Although many noncrystalline materials behave viscously at high temperatures, the curve of viscosity versus temperature does not continue as an extrapolation of Figure 9.4 to lower temperatures would suggest; rather, the curve bends over to the ordinate and intersects it at room temperature at about $10^{20}$ poises rather than at the extrapolated value of $10^{60}$ poises, as shown in Figure 9.5. It can be seen that the viscosity changes very rapidly in a relatively small temperature range; the temperature at which the change is most rapid is called the *glass transition temperature,* abbreviated $T_g$.

The curve in Figure 9.5 bends over because at low temperatures the deformation is not purely viscous, that is, $\eta$ is not a coefficient of *ideal,* or *Newtonian, viscosity.* In principle, at least, models of such *viscoelastic* behavior can be devised. The simplest are merely combinations of a dashpot (representing pure viscous flow) and a spring (ideal elastic deformation) and were proposed by Maxwell and Voigt (Figure 9.6). By varying the magnitudes of the spring constants and the dashpot viscosities, it is possible to start

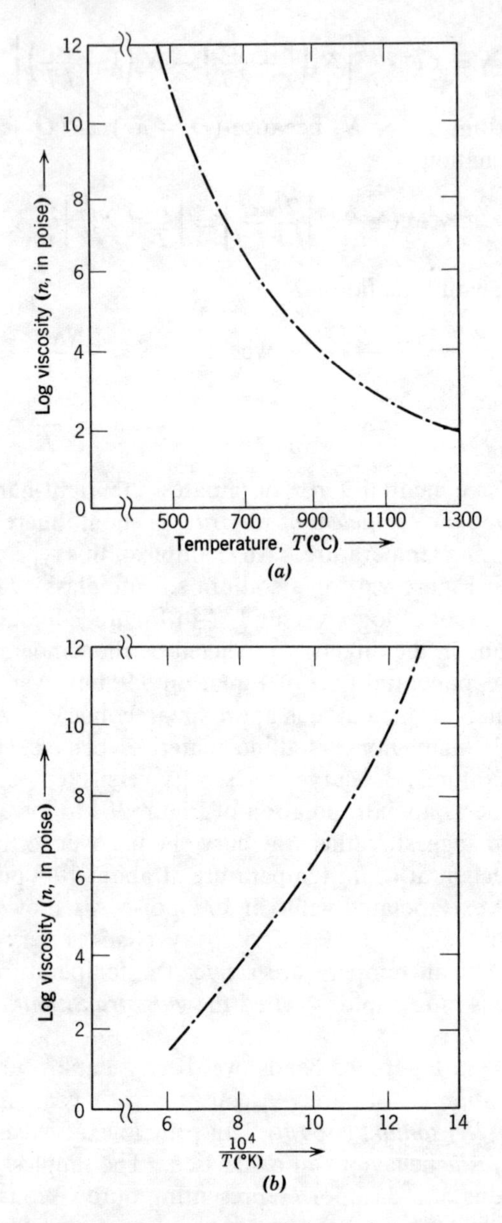


Figure 9.4 Experimentally determined coefficients of viscosity, $\eta$, as a function of temperature for a sodium silicate glass.

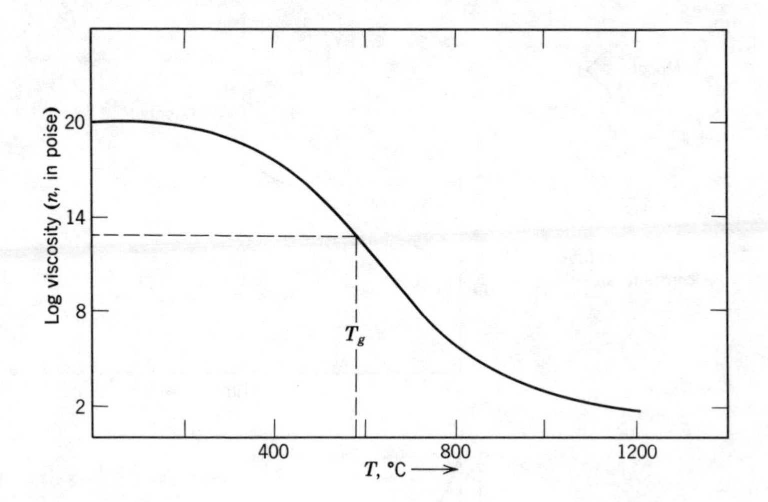


Figure 9.5 The logarithm of the coefficient of viscosity of a sodium silicate glass from 0°C to 1300°C.

either with almost pure elastic deformation (representing very low-temperature deformation) or with almost pure viscous flow (representing very high-temperature deformation); whether a modulus $E$ or a coefficient of viscosity $\eta$ is measured depends on the temperature, the duration of the experiment, and the type of measurement.

In the melting range, the viscosity of glass is in the range of 50 to 100 poises. It can be formed by such procedures as blowing, rolling, or drawing when the viscosity is $10^4$ to $10^8$ poises. Internal stresses can be substantially relieved in a period of 15 minutes at a temperature called the "annealing point," at which the viscosity is $10^{13.4}$ poises. At low temperatures, where viscous flow is improbable and plastic flow impossible, glasses behave elastically. Glasses are extremely notch sensitive at these temperatures, and, as was discussed in Chapter 7, unless great care is taken to remove internal and surface microcracks, their tensile strengths are unpredictably low, although their compressive strengths may be quite high.

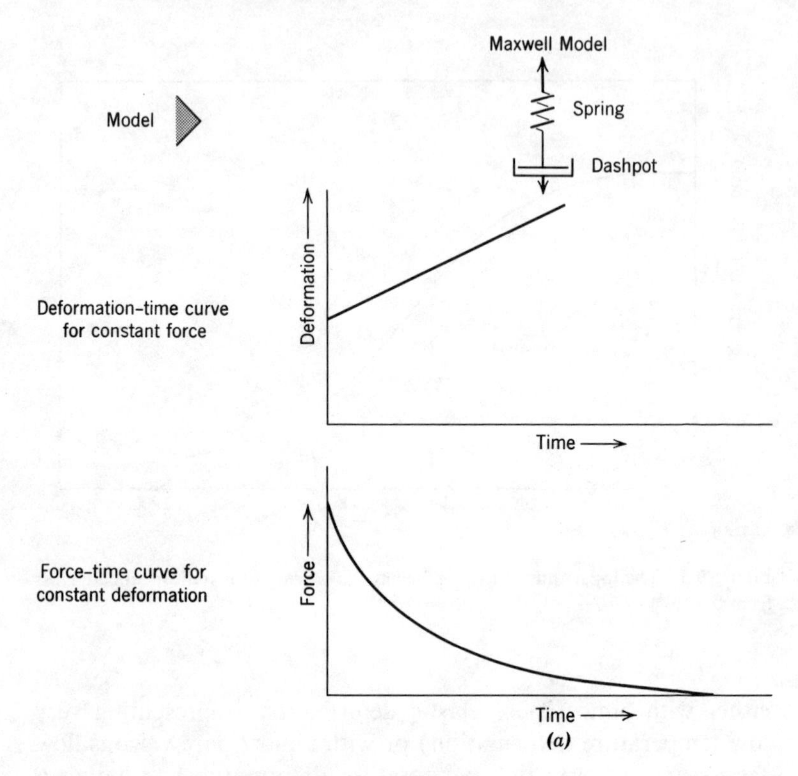


Figure 9.6 (*a*) The deformation-time and force-time curves for a Maxwell model. (*b*) The deformation-time and force-time curves for a Voigt model.

## 9.4 FABRICATION OF CERAMIC BODIES

Most ceramic bodies are produced either directly from the melt as a glass or by the consolidation of powdered ceramic materials. In powder fabrication, the powders are compacted by mechanical pressing or by slip casting; the latter is the name of a process in which a suspension of powders in a liquid is cast in a porous mold. The liquid is absorbed in the mold, leaving a solid cast powder mass which has some strength. The residual porosity in both cases is reduced by sintering at high temperatures. In certain situations hot-pressing, a combination of both the pressing and sintering operations, may be employed.

In sintering, all the components may be solids or some may be

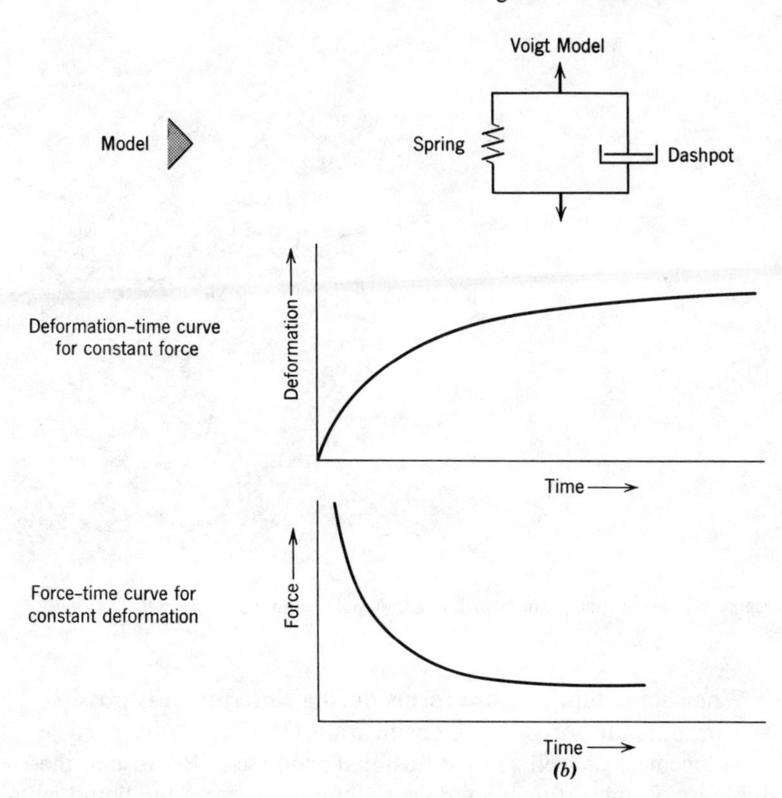

*(b)*

liquid, depending on the material being treated. When they are all solid, porosity is reduced through diffusional mass transport processes which change both the shape and size of the pores. It is almost impossible, however, to eliminate all porosity. In most metals, any porosity remaining after sintering can be removed by a combination of cold-working and annealing or by hot-working. However, since ceramics can be neither cold-worked nor hot-worked, these methods cannot be employed. If ceramics are sintered for a long time, porosity does continue to decrease at a slow rate after the initial rapid amount of densification. But during this long sintering procedure, abnormal grain growth, where a few grains begin growing much more rapidly than most other grains, occurs (see Figure 9.7). This can have a serious effect on the strength.

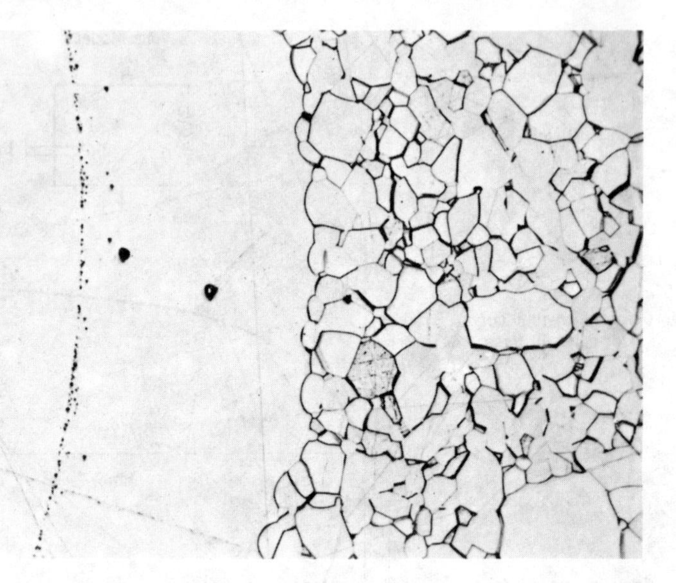

Figure 9.7 Abnormal grain growth in an alumina ceramic (courtesy R. L. Coble).

When some liquid phase forms during sintering, it is possible to eliminate all porosity. Densification then occurs by particle rearrangement as well as by diffusional processes. Because of the difficulty of nucleating a solid crystalline phase from the liquid in ceramic materials, the microstructure resulting from liquid-phase sintering is usually a solid crystalline phase embedded in a glassy matrix.

## 9.5 STRENGTH OF CERAMICS

As mentioned previously, most ceramic materials are brittle. Even in those few materials of NaCl structure which are ductile as single crystals, the effects of impurity locking of dislocations, the extreme temperature and strain-rate sensitivity of the yield stress (which produce notch sensitivity and ductile-to-brittle transitions), and the limited number of independent slip systems cause the randomly oriented polycrystalline forms of these materials to be brittle. Furthermore, as mentioned above, sintered solids invariably contain pores; these act as stress concentrators where brittle fracture intrudes. Even when porosity is eliminated by liquid-phase sintering, the inevitable presence of a continuous glassy phase leads to brittleness.

Because of notch sensitivity and the extreme difficulty of plastic or viscous deformation at low temperatures, the strengths of ceramics under a tensile load are low and generally unpredictable. However, most of these materials exhibit high compressive strengths and the designer has learned to use them to sustain appreciable loads.

The high temperature creep observed in a number of ceramic materials results from the thermal activation of secondary slip systems and from such creep processes as dislocation climb, stress-induced diffusion, and grain boundary sliding. As a rule, however, a ceramic material has a far greater resistance to creep than a metal of similar melting point.

## 9.6 THERMAL CONDUCTIVITY

One of the chief uses of inorganic nonmetallics is as thermal insulators. Because these materials have either ionic or covalent bonds, there are not enough free electrons to bring about electronic thermal conductivity. Instead, heat is conducted by *phonon* conductivity, the interaction of anharmonic lattice vibrations, and, at high temperatures, by radiant heat transfer.

The thermal conductivity of ordinary ceramics can be altered by several factors. It is generally lower than that of single crystals, because grain boundaries scatter radiant energy and thus lower the effective radiant heat conductivity. Similarly, at low temperatures, any porosity, as in insulating brick, decreases the conductivity, as is shown in Figure 9.8. At high temperatures, however, large pores and higher emissivities can raise conductivity, as radiant heat conduction is directly proportional to the pore size. With decreasing crystallinity, the probability of phonon scattering increases, making heat conduction more difficult. Figure 9.9 shows the effect of structure on the thermal conductivity of $SiO_2$. In these experiments, quartz crystals were subjected to neutron irradiation; increasing degrees of radiation disrupt the crystal structure, until finally a glassy structure is created.

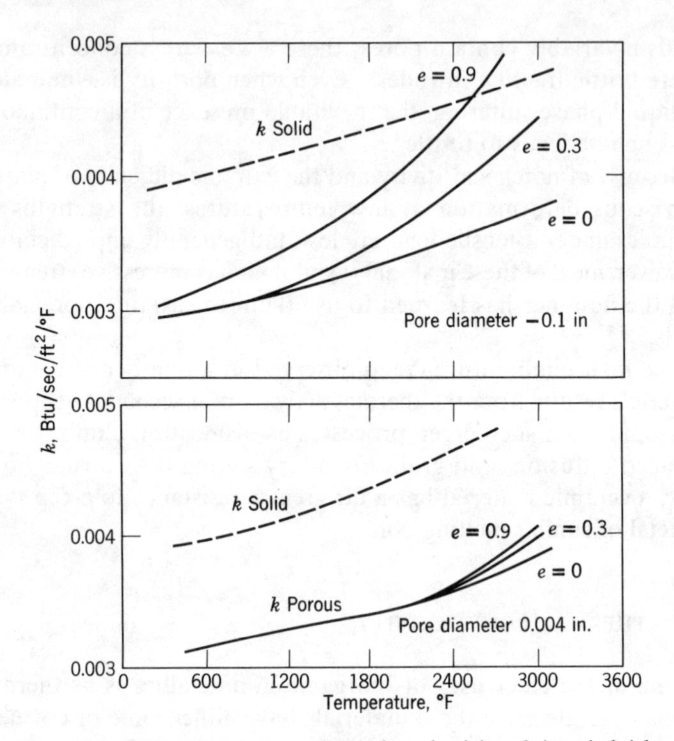


Figure 9.8 Effect of porosity on thermal conductivity of zirconia brick.

## 9.7 THERMAL STRESSES

Thermal stresses are developed in ceramic materials in which expansion or contraction caused by changes in temperature are mechanically constrained. In recent years glass articles with improved strengths have been produced by heat treatments which relieve thermal stresses. In metal articles enameled with a ceramic glaze, sufficient stresses can develop in the enamel on cooling because of the difference in thermal expansion coefficients, to cause cracking (crazing). A similar problem is that of *thermal shock:* the phenomenon which causes a fragile china cup to break when a hot liquid is suddenly poured into it. Thermal shock can be related to several basic properties common to ceramic materials. Because of their low thermal conductivity, these materials do not heat or cool uniformly when subjected to a sudden temperature change, and therefore thermal stresses are developed within them which cannot be relieved by deformation processes. If surface flaws are present in the material, these thermal stresses are sufficient to cause failure. The maximum rapid temperature change which an infinite ceramic slab subject to thermal shock can withstand is given by the equation:

$$\Delta T = \frac{\sigma_f(1 - \nu)}{E\alpha} \tag{9.10}$$

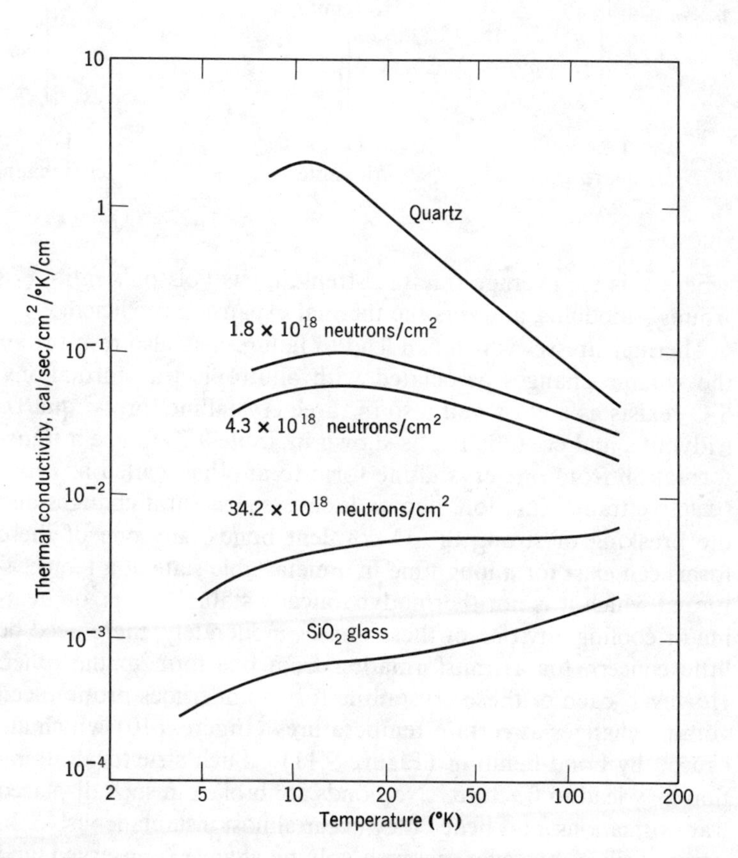


Figure 9.9 Effect on neutron irradiation on the thermal conductivity of a quartz crystal (from R. Berman, F. E. Simon, P. G. Klemens, and T. M. Fry, *Nature,* 166, 1950, p. 277).

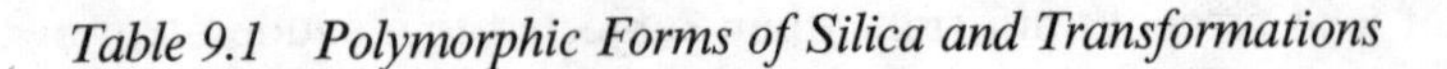

*Table 9.1 Polymorphic Forms of Silica and Transformations*

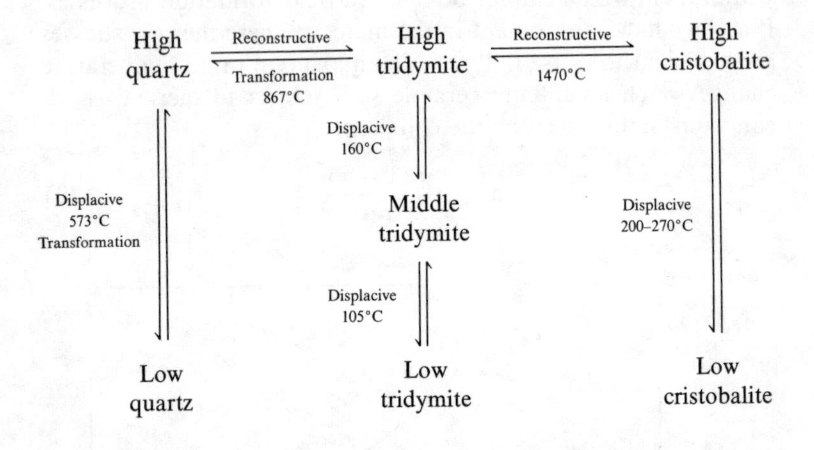


where $\sigma_f$ is the average fracture strength, $\nu$ is Poisson's ratio, $E$ is Young's modulus, and $\alpha$ is the thermal expansion coefficient.

Thermal stresses which can lead to failure may also result from the volume changes associated with allotropic transformations. $SiO_2$ exists as a glass and also in three crystalline forms: quartz, tridymite, and cristobalite, as shown in Table 9.1. Since a transformation from one crystalline form to another, called a reconstructive transformation, demands severe structural changes and the breaking of strong Si—O covalent bonds, any one of these forms can exist for a long time in a metastable state at a temperature at which it is not thermodynamically stable. Thus, on heating or cooling any one of these phases moderately, there need be little concern for a transformation from one form to the other. However, each of these crystalline forms undergoes pronounced volume changes at certain temperatures (Figure 9.10) which are caused by bond-bending (Figure 9.11). Such structural alteration may lead to fracture. No bonds are broken in such displaced transformations, and hence they occur almost instantaneously. In vitreous silica, however, no such volume change is observed, and thus, if silica is to be used over a wide range of temperatures, the vitreous form must be employed.

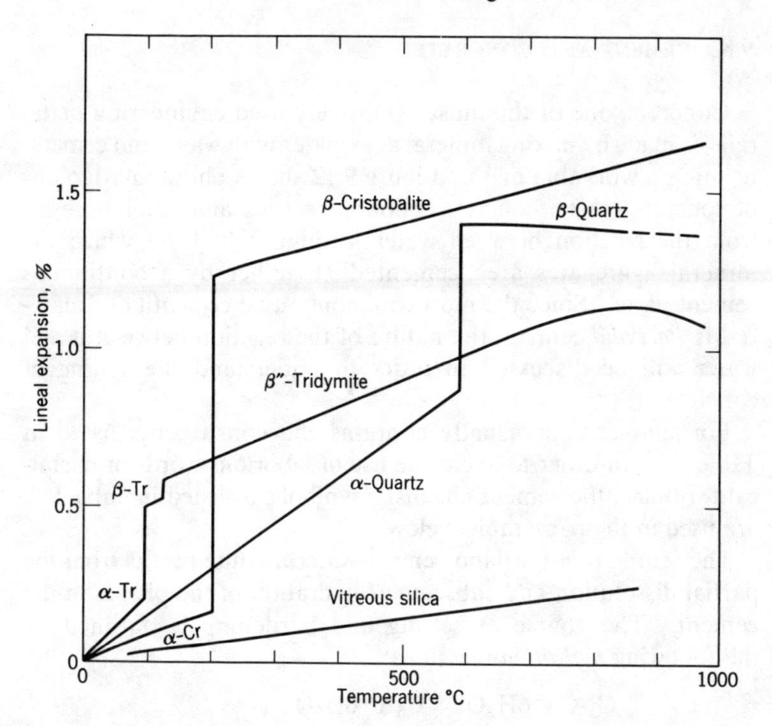


Figure 9.10 Thermal expansion of $SiO_2$ phases (From L. H. Van Vlack, *Elements of Materials Science,* Addison-Wesley, 1959, p. 127).

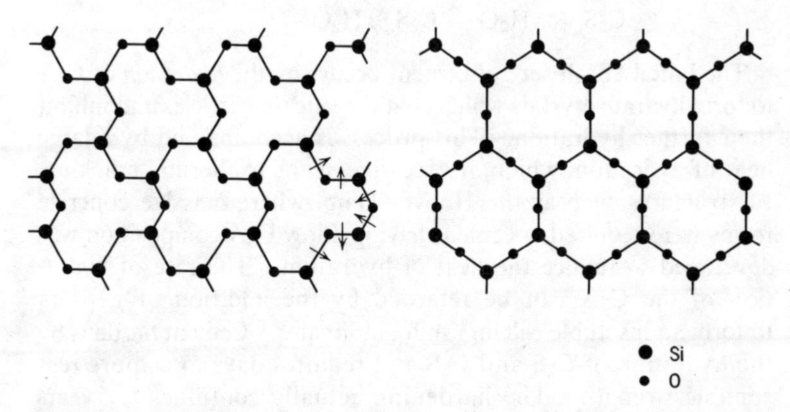


Figure 9.11 Bond-straightening, producing instantaneous volume expansion, as in $\alpha$ quartz → $\beta$ quartz. (From Van Vlack, *op.* cit, p. 126).

## 9.8 CEMENT AND CONCRETE

Concrete, one of the most extensively used engineering materials, is made by mixing mineral aggregates with water and cement, forming a workable mass. Figure 9.12 shows photomicrographs of concrete. After concrete is poured, setting and bonding result from the reaction between water forming a body in which the mineral aggregates are "cemented" together by a continuous cement paste. Since the most commonly used cementitious material is *Portland* cement, the nature of the reaction between it and water will be discussed in order to understand the setting of concrete.

Portland cement usually contains the constituents listed in Table 9.2. In order to avoid the use of laborious words or chemical formulae, the cement chemist's symbols, as listed in Table 9.2, are used in the paragraphs below.

The setting of a Portland cement-water mixture results from the partial dissolution and subsequent hydration of the phases in the cement. The course of setting and hardening is outlined by the following *hydration* reactions:

$$C_3A + 6H_2O \rightarrow C_3A \cdot 6H_2O$$

$$C_4AF + 7H_2O \rightarrow C_3A \cdot 6H_2O + CF \cdot H_2O$$

$$C_3S + xH_2O \rightarrow C_2S \cdot xH_2O + Ca(OH)_2$$

$$C_2S + xH_2O \rightarrow C_2S \cdot xH_2O$$

The initial "flash set" of cement occurs by the *hydration* of $C_3A$ to form hydrate crystals which coat the silicate particles and inhibit their further hydration. This process is accompanied by a large heat of hydration, which, if excessive, leads to thermal cracking. In structures such as the Hoover Dam, where massive concrete forms were required, a cement having a low $C_3A$ composition was developed to reduce the heat of hydration. The rate of hydration of the $C_3A$ can be retarded by the addition of gypsum to form an insoluble calcium sulfoaluminate*. Cement hardens by the hydration of $C_2S$ and $C_3S$ and requires days to acquire reasonable strength. The hardening actually continues for years and is thought to be due to the formation of *tobermorite* gel ($C_3S_2 \cdot 3H_2O$), a colloidal material in which the calcium silicates

* $3CaO \cdot Al_2O_3 \cdot xCaSO_4 \cdot yH_2O$ where $x = 1$ to 3 and $y = 10.6$ to 32.6.

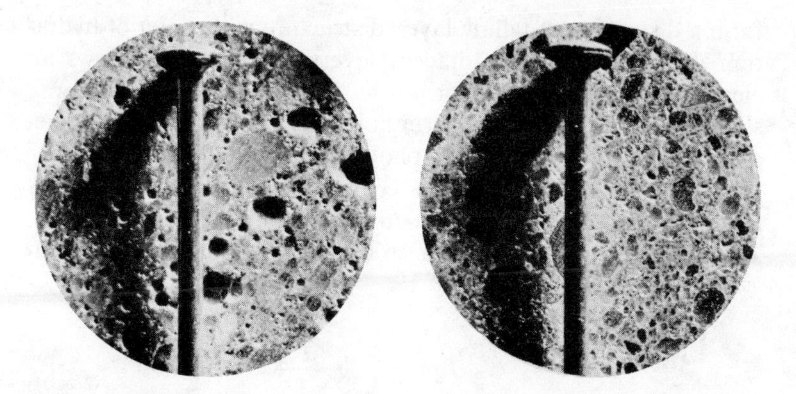

Figure 9.12 An ordinary straight pin photographed on two cross-sectional specimens of concrete. On the left is a specimen of air-entrained concrete showing air cells, which total billions per cubic foot. On the right is a section of nonair-entrained concrete (courtesy Portland Cement Association).

*Table 9.2 Constituents of Portland Cement*

| CONSTITUENT | CEMENT CHEMIST'S SYMBOL | WEIGHT PERCENT IN TYPE 2 PORTLAND CEMENT |
|---|---|---|
| Tricalcium silicate— $3CaO \cdot SiO_2$ | $C_3S$ | 46 |
| Dicalcium silicate— $2CaO \cdot SiO_2$ | $C_2S$ | 28 |
| Tricalcium aluminate— $3CaO \cdot Al_2O_3$ | $C_3A$ | 11 |
| Tetracalcium aluminoferrite— $4CaO \cdot Al_2O_3 \cdot Fe_2O_3$ | $C_4AF$ | 8 |
| Gypsum—$CaSO_4 \cdot 2H_2O$ | . . . | 3 |
| Magnesia—MgO | . . . | 3 |
| Calcium oxide—CaO | C | 0.5 |
| Sodium oxide and potassium oxide— $Na_2O$, $K_2O$ | | 0.5 |

form a distorted crystalline layered structure with water of hydration dissolved between adjacent layers. Figure 9.13 shows an electron micrograph of tobermorite. In cement, the tobermorite sheets appear to roll up to form hollow fibers, which form aggregates as shown in the electron photomicrograph of Fig. 9.13*b*.

Studies of the hardening of cement have indicated that $C_3S$ requires about 30 days to reach 70 percent of its ultimate strength. The hydration of $C_2S$ proceeds so slowly that it only reaches two-

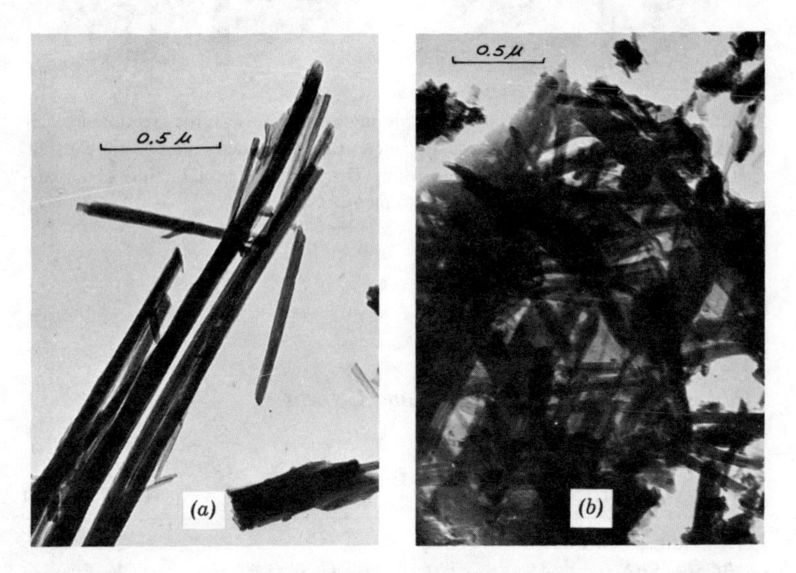


Figure 9.13 Tobermorite gel (from S. Brunauer, *American Scientist*, 50, March 1962).

thirds of its final strength in 6 months. Setting can be hastened by using finer cement and raising the temperature (cement will set in one-third the time at 100°F that it takes at 40°F), as well as by controlling composition. When determining the composition of a cement, the amounts of free calcium and magnesium oxide must be minimized because both these oxides undergo a large volume expansion when they hydrate. This can cause cracking and actual disintegration of the hardened cement.

In the preparation of a concrete mix, it is necessary to employ size distributions of mineral aggregates which give a high packing density. This not only reduces the amount of cement required but also results in a stronger, less permeable, and, hence, more durable product.

Water is added to concrete mixes to produce hydrates and to lubricate the aggregates and make the mix more workable. However, porosity may be produced in the concrete either from air

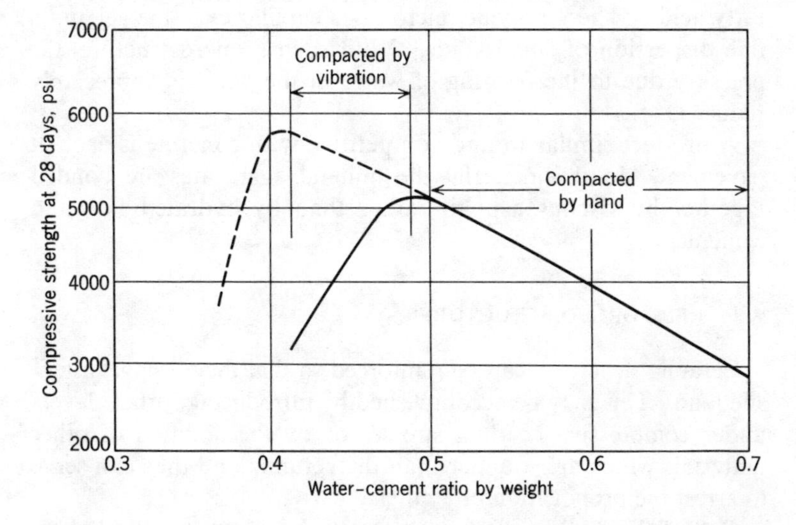


Figure 9.14 The effect of the water-cement ratio on the strength of concrete for different methods of compacting (from Z. D. Jastrzebski, *Nature and Properties of Engineering Materials,* John Wiley and Sons, New York, 1959, p. 501). For concretes of low water-cement ratios the shape of the dashed curve can be different if trapped air is removed.

trapped in the mix or from excess water. If the water-cement ratio in a mix is too low, the trapped air cannot be readily removed, and the resulting porosity decreases the strength. If, however, the water-cement ratio is too high, the porosity resulting from the eventual evaporation of the excess water also affects the strength adversely. Thus, as is shown in Figure 9.14, there are intermediate *water-cement ratios* which produce optimum strengths, for different methods of compaction.

Hardened Portland cement is not an especially dense material. Although less permeable to water than some stones and rocks, it nevertheless contains capillary pores as well as statistical voids in its gel structure. The presence of water in such a structure can, during freezing weather, lead to an increase in internal pressure which may cause local fracture of the gel structure. With alternate thawing and freezing, such damage can become progressive. To reduce this tendency it has been found useful to employ addition agents to the concrete mix such as oils, waxes and salts of the fatty acids. These provide micron size capillaries in the gel and a fine dispersion of small voids. With such a microstructure, the pressure due to the freezing of water in the voids is appreciably reduced.

A product similar to and competitive with concrete is asphalt pavement. In this material the mineral aggregates are bonded together by viscous asphalt rather than by hydrated Portland cement.

## 9.9 REINFORCED STRUCTURES

Ceramic structures can be reinforced so that they support a tensile load. This may be accomplished by introducing surface layers under compressive residual stresses or by the addition of other materials which are tougher than the ceramic and thus can serve to arrest the propogation of a crack.

One of the earliest reinforced materials was "tempered" glass. By a process of heating and then cooling, compressive stresses are introduced into the surface of the glass. Any applied tensile force must exceed the compressive stress before there is a net tensile stress on the surface of the glass. Thus the material can withstand appreciable tensile stress before the tensile stress at the surface becomes sufficient to propagate cracks and cause failure. Glass is tempered by heating it to a temperature above the softening point and then quenching it with jets of air. On quenching, the surface contracts more quickly than the interior, but the interior is still hot enough to flow and adapt to the changes in the surface. As the interior cools, it also tends to contract; however, since the surface is cooler than the interior, it cannot adjust itself to any volume changes in the interior. As a result of this proce-

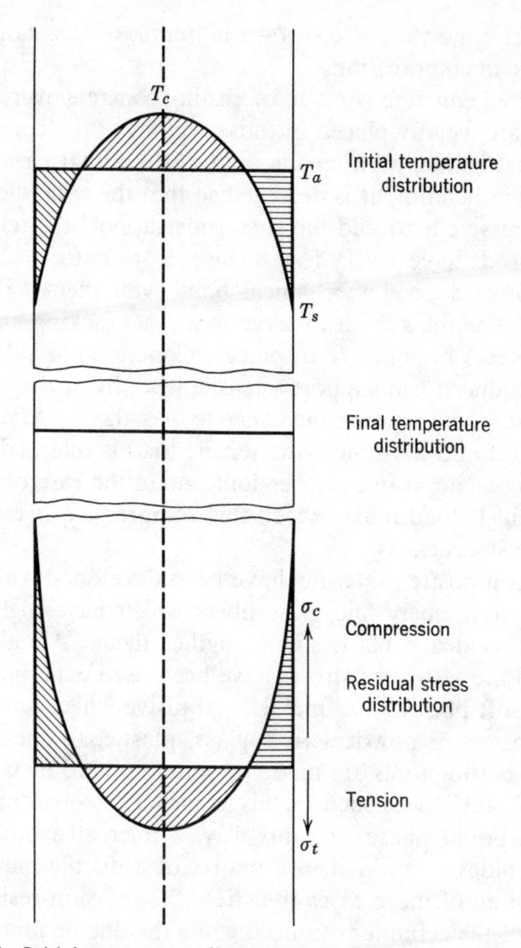


Figure 9.15 Initial temperature distribution, final temperature distribution, and resulting residual stresses in tempered glass (from Kingery, *Introduction to Ceramics,* John Wiley and Sons, New York, 1960, p. 640).

dure, the surface is placed in compression and the interior in tension, as shown in the stress profile in Figure 9.15. Different methods have been devised for creating compressive stress in glass surfaces, one of which is to coat a glass having a high coefficient of thermal expansion with one having a low coefficient. On cooling, the material on the interior, having the high coefficient, tends to contract more than the surface material, which, again, places the surface in compression.

Reinforced concrete is made by casting concrete over steel rods. The rods are usually placed in those regions of the concrete form which must withstand a tensile load. If the final structure is to be loaded in bending, it is designed so that the concrete supports the compressive load and the steel rods support the tensile load. The steel rods have a very rough surface, so that the concrete, in setting, forms a good mechanical bond with them. This rough surface also enables them to serve to arrest cracking and to hold cracked pieces of concrete in place. Concrete may also be prestressed so that it can support a tensile load, by putting steel rods under tension and pouring the concrete over them. After the concrete has set and hardened, the tensile load is released from the rods, thus placing it in compression. As in the case of tempered glass, a tensile load must exceed this compressive stress before a net tensile stress exists.

Many composite materials have been developed which utilize ceramics. Extremely fine glass fibers which have high strength have been bonded in plastics to strengthen them. Alumina whiskers, which are also very strong have been used experimentally to reinforce soft plastics and metals. Abrasive wheels are made by bonding abrasive powders in a glass, plastic, or metal matrix. "Cermet" cutting tools are made by sintering hard metal carbides (WC, TiC, TaC) with such metals as nickel or cobalt in the presence of a liquid phase. In this way, a microstructure of hard, brittle carbides embedded in a matrix of a ductile metal is produced. In all of these cases the strength, abrasion-resistance, or hardness of the ceramic is utilized while the ductile matrix arrests crack growth and thus toughens the composite material.

## 9.10 GRAPHITE

A nonmetallic material that is ever finding new applications is graphite. It is usually made from a mixture of coke and pitch, cold-formed by pressing or extrusion and then heat treated to develop a sintered structure. It can also be hot-formed. The heat-treated article consists of grains of crystallized materials in a matrix of finer grained material which itself is neither fully crystalline nor wholly amorphous. The porosity and degree of crystallinity achieved in manufacture depend to a great extent on the raw material used.

A singular type of graphite from the standpoint of microstructure is pyrolytic graphite. It is formed by passing hot $CH_4$ gas over a heated surface. Crystals of graphite deposit with their *c* axis normal to the underlying surface in parallel bundels of like-oriented crystallites. This type of graphite is the most anisotropic of the various kinds produced (see Volume I). The crystallographic anisotropy is responsible for its anisotropy in thermal conductivity and mechanical properties. Such anisotropies can lead to complicated stress and corrosion patterns as well as to dimensional instability in high temperature applications where cooling of one side is practiced. In spite of such shortcomings in pyrolytic and other kinds of graphite, engineers have learned how to specify and to use it for many purposes.

As a lubricant in metal-forming processes or in bearing material, graphite provides a low coefficient of friction. It does so only in atmospheres containing oxygen or compounds thereof. At high enough temperatures it can actually raise the coefficient of friction between sliding surfaces. In vacuum or in an inert gas it is very high even at room temperature. Catastrophic wear takes place in vacuum when graphite rubs against steel or copper. Porous polymeric or metal-bonded mixtures of graphite metal and ceramic powders have been used after pressing and sintering (or curing) as friction clutches or brakes. The opposing member of the friction pair may be a similar composite or a solid steel or cast iron.

Other nonmetallic materials which have basal cleavage planes analogous to graphite are molybdenum disulfide, talc, and boron nitride. A synthetic high pressure modification of boron nitride possesses a diamond cubic rather than a sheet structure. Its properties consequently simulate those of diamond.

## DEFINITIONS

*Ceramics.* Inorganic, nonmetallic materials. They are generally ionic or covalent chemical compounds.

*Cementation.* The process by which individual particles are bonded together by the hardening of a binder phase which "cements" these particles, forming a rigid body.

*Gel.* A colloidal material consisting of a fine-scale mixture of solids, liquids, and frequently gases. Gels form by the coagulation of a liquid and are characterized by the fact that the solid material in suspension does not settle out over long periods of time.

*Viscous.* That type of deformation behavior under which a body cannot support an applied shear stress. Viscosity is that property of a viscous material which measures its resistance to flow; thus highly fluid materials have low viscosities.

*Poise.* The unit of viscosity, measured in dyne $sec/cm^2$.

*Thermal Stress.* A stress in a body over which there is a thermal gradient; the stress results from the differences in the amounts of thermal expansion (or contraction) in regions of the body at different temperatures or with different coefficients of expansion.

*Thermal Shock.* The sudden failure of a body subjected to a rapid temperature change; it occurs most readily in brittle materials which have high coefficients of thermal expansion and low thermal conductivities.

*Hydration.* The reaction of a compound with water which forms a hydrate of that compound; the reaction is characterized by the formula: $AB + xH_2O \rightarrow AB \cdot x(H_2O)$.

## BIBLIOGRAPHY

INTRODUCTORY READING:

Hauth, W. T., "Crystal Chemistry in Ceramics", *Bull. Amer. Cer. Soc.* **30** (1951), Parts 3, 5, 6, 7, and 8.

National Bureau of Standards, "Mechanical Behavior of Crystalline Solids," *National Bureau of Standards Monograph* 59, 1963.

Van Vlack, L. H., *Elements of Materials Science,* Addison Wesley, Reading, Mass., 1964, 2nd ed., Chapter 8.

Van Vlack, L. H., *Physical Ceramics for Engineers,* Addison, Wesley, Reading, Mass., 1964.

SUPPLEMENTARY READING:

Cottrell, A. H., *The Mechanical Properties of Matter,* John Wiley and Sons, New York, 1964, Chapters 3, 7, 9, 10, 11.

Kingery, W. D., *Introduction to Ceramics,* John Wiley and Sons, New York, 1960, Chapters 16, 17.

Gilman, J. J., *Mechanical Behavior of Ionic Crystals in Progress in Ceramic Science,* Vol. I, Pergamon Press, New York, 1961.

ADVANCED READING:

Brunauer, S., "Tobermorite Gel," *American Scientist* **50,** (March, 1962), p. 210.

Bogue, R. H., *Chemistry of Portland Cement,* Rheinhold Publishing, New York, 1955.

Eitel, W., *The Physical Chemistry of the Silicates:* University of Chicago Press, Chicago, 1954.

Kronberg, M. L., "Plastic Deformation of Single Crystals of Sapphire, Basal Slip and Twinning," *Acta. Metallurgica,* **5** (Sept., 1957), p. 517.
Wooster, W. A., *A Textbook on Crystal Physics,* Cambridge University Press, Cambridge, 1938.

PROBLEMS

9.1 List the conditions necessary for a crystalline oxide to be an easy glass former.

9.2 Account for the fact that a sharp knife placed at the top of a cleavage plane in a well-supported crystal need only be struck lightly by a hammer to cleave it.

9.3 (a) Why do cubic ionic crystals such as MgO, NaCl, and LiF cleave on cube planes?
(b) On what planes do BCC transition metals cleave?
(c) Do FCC metals cleave?

9.4 (a) Explain why wood is as strong as it is.
(b) Use a similar argument to describe resin-bonded glass fibers.
(c) How would you achieve three-dimensional strengths in fiber composites?
(d) Why are cobalt-bonded tungsten carbide cutting tools superior to cast tungsten carbide? (What percentage of the tool is cobalt?)

9.5 Defining the strains produced by a given amount of plastic deformation on a given slip system as follows:

$$\epsilon_x = \alpha\beta_x n_x; \quad \epsilon_y = \alpha\beta_y n_y; \quad \epsilon_z = \alpha\beta_z n_z$$

$$\epsilon_{xy} = \frac{\alpha}{2}(\beta_x n_y + \beta_y n_x)$$

$$\epsilon_{yz} = \frac{\alpha}{2}(\beta_y n_z + \beta_z n_y)$$

$$\epsilon_{xz} = \frac{\alpha}{2}(\beta_z n_x + \beta_x n_z)$$

where $\alpha$ equals the scalar strain parameter, $\beta_i$ equals the $i$ component of the slip direction, and $n_i$ is the $i$ component of the slip plane normal. Show that the comparable amounts of deformation on the slip system $(\bar{1}10)$ [110] in a rock salt material produce exactly the same strains as on the orthogonal slip system $(1\bar{1}0)$ [110].

9.6 By the same method as outlined in Problem 9.5, show that the sum of the strains produced on the nonorthogonal slip systems (110) $[1\bar{1}0]$ and (101) $[10\bar{1}]$ are equal to the strains produced on the third nonorthogonal system (011) [011].

9.7 Why are metal oxides notably more brittle than the metals themselves, even though they often have equivalent dislocation densities?

9.8 At room temperature, polycrystalline AgCl can deform much like a ductile metal. At liquid nitrogen temperatures, however, it behaves in a brittle fashion. Discuss what is the single most important factor leading to this behavior.

9.9 Molten glass can be blown into fibers. Why is it impossible to produce $Al_2O_3$ fibers in this manner?

9.10 For a sintered material whose theoretical density is known, outline a technique by which the volume percentage of total porosity may be determined. Total porosity includes both parts which are closed to the surface and those which are open to the surface.

9.11 What is the maximum rapid temperature change a glass article with surface cracks of maximum length 1 $\mu$ can withstand without failing catastrophically? $E = 7 \times 10^{11}$ dynes/cm²; $\gamma = 300$ ergs/cm²; $\nu = 0.25$; $\alpha = 8 \times 10^{-6}$ cm/cm °C.

9.12 Discuss the relative merits of utilizing a structural member made of molybdenum, crystalline quartz, or fused quartz in a furnace which cycles between 1200°C and room temperature every 24 hours.

9.13 The packing factor (percent of space filled) of a gravel-sand mixture containing 25 percent sand and 75 percent gravel is 90 percent for a particular type of gravel and sand, both of which consist of uniform-size particles. How could this packing factor be increased? Why is it not economically desirable to attain high packing factors?

9.14 Compare the mechanism of strengthening that accompanies the setting of cement in concrete with that which occurs in the sintering of ceramic articles fabricated from powders.

9.15 Discuss the seeming contradiction between the two facts that some glass articles can be strengthened by annealing to remove residual stresses, whereas other glass articles are strengthened by tempering, an operation which deliberately introduces residual stresses.

9.16 Glass beads have long been made by quenching drops of molten glass in water. The beads have a teardrop shape with long thin tails at one end. The body of the teardrop can withstand the impact of a hammer. However, if the tail is broken by hand, the whole bead will shatter into many pieces. Explain these observations.

9.17 In the manufacture of tempered glass, why must the article first be heated above its softening temperature before being quenched? What will happen if it is quenched before it is heated to a high enough temperature?

9.18 (a) Indicate on a sketch the initial temperature distribution, the final temperature distribution, and the resulting residual stress distribution in a rod of tempered glass.

(b) Indicate on a sketch the stress distribution when a tempered glass rod is pulled in tension.

(c) Compare the stress distribution in prestressed concrete both when unloaded and loaded in tension with those in tempered glass.

9.19 One of the current ideas of how to put compressive residual stresses on the surface of glasses is by the exchange of small ions for large ions. Estimate the maximum stress that one might achieve in a glass containing 0.15 mole fraction $Na_2O$ if half the sodium ions on the surface are exchanged with potassium ions.

Partial molar volume $= V_A = (\partial V/\partial m_A)_{T,P,m_B,m_c}$
$\overline{V}_{Na_2O} = 25\ cm^3/mole$
$\overline{V}_{K_2O} = 27\ cm^3/mole$
Molar density of glass $= 0.04\ cm^3/mole$
Elastic modulus $= 10^7$ psi
Poisson's ratio $= 0.25$

CHAPTER TEN

# *Polymers*

The mechanical behavior of polymers is markedly influenced by molecular structure. The degree of polymerization, branching, and cross-linking also affects strength. Increasing the crystallinity of polymers can similarly increase strength and density. As polymers are heated they pass through five general states: glassy, leathery, rubbery, viscous-rubbery, and liquid. This behavior has been explained with the aid of a viscoelastic model of molecular movements. Polymeric materials are sometimes classified as thermoplastic or thermosetting, depending on their behavior at elevated temperatures. Elastomers are polymers distinguished by great reversible extensibility and unique thermal behavior.

## 10.1 INTRODUCTION

Polymers, as pointed out in Volume I, consist of long molecules made up of relatively simple molecular units. The most common polymers are those made from compounds of carbon, but polymers can also be made from inorganic chemicals, such as silicates and silicones. The mechanical properties of organic polymers are highly sensitive to molecular configuration, which is itself sensitive to mode of manufacture. An understanding of elastic modulus, deformation under stress, tensile strength, yield strength, and impact behavior of polymers therefore requires the simultaneous consideration of composition and mode of manufacture. Although more emphasis has been placed on elastic modulus and tensile strength in the text, other mechanical properties of polymers are also structure sensitive.

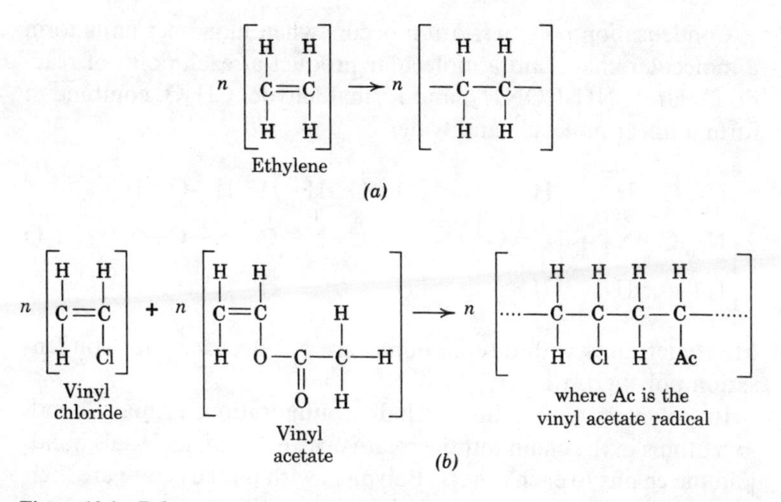


Figure 10.1 Polymerization reactions: (*a*) polymerization of ethylene, (*b*) copolymerization of vinyl chloride and vinyl acetate.

## 10.2 TYPES OF POLYMERS

Classification of polymers according to mode of manufacture results from the fact that long-chain molecules are produced by two general types of reactions, namely, *addition* and *condensation* polymerization. Addition polymerization occurs when one or more types of monomers become attached end-to-end to form a chain. The polymerization of the unsaturated hydrocarbon ethylene, $CH_2{=}CH_2$, to form polyethylene, is the simplest example of this reaction (Figure 10.1*a*). Similarly, vinyl chloride and vinyl acetate may be polymerized together to form a copolymer (see Figure 10.1*b*).

Monomers containing double bonds may be *bifunctional,* that is, the double bond may "open" and form two sites for molecular bonding. The polymerization of such monomers therefore produces saturated chain molecules between which primary bonds cannot be formed. The presence of relatively large radicals on the monomer which form side groups or branches on the resultant molecular chain also influence properties.

Condensation polymerization occurs when monomer units form a molecular chain and a molecular product at each point of reaction. Urea, $NH_2CONH_2$, and formaldehyde, $CH_2O$, combine to form a linear molecule and water

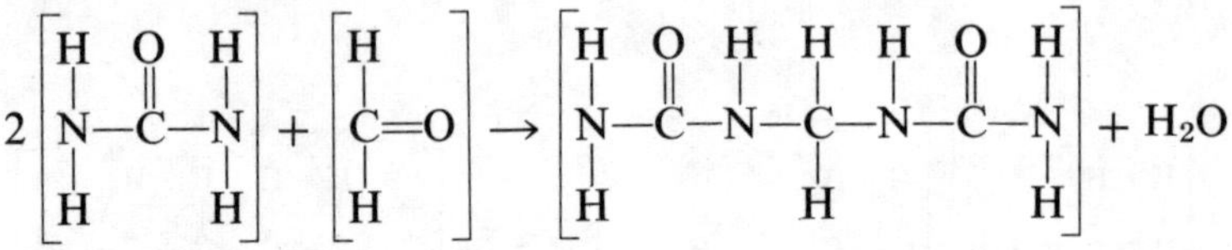

$$2\left[\begin{array}{ccccc} H & & O & & H \\ | & & \| & & | \\ N & - & C & - & N \\ | & & & & | \\ H & & & & H \end{array}\right] + \left[\begin{array}{ccc} H & & \\ | & & \\ C & = & O \\ | & & \\ H & & \end{array}\right] \rightarrow \left[\begin{array}{ccccccccccccc} H & & O & & H & & H & & H & & O & & H \\ | & & \| & & | & & | & & | & & \| & & | \\ N & - & C & - & N & - & C & - & N & - & C & - & N \\ | & & & & & & | & & & & & & | \\ H & & & & & & H & & & & & & H \end{array}\right] + H_2O$$

Monomer units with double bonds are not necessary for condensation polymerization.

In polymers with a linear chain configuration, primary bonds join atoms in the chain but the relatively weak van der Waals bonds join the chains to each other. Polymers with this structure are thermoplastic. They soften with increasing temperatures and are readily deformed. On cooling, they assume their original low-temperature properties but retain the shape into which they were molded.

When covalent cross-links or strong hydrogen bonds join the chains of a polymer together, as in cellulose, an increase in temperature does not facilitate plastic deformation. Such polymers are called thermosetting; they remain relatively strong until chemical decomposition sets in. They are actually made of three-dimensional networks of covalent bonds instead of cross-linked chain molecules. Bakelite is a thermosetting network polymer formed by the condensation polymerization of phenol and formaldehyde; its structure is shown in Figure 10.2.

## 10.3 CRYSTALLINITY OF POLYMERS

Polymers, like other noncrystalline solids, exhibit short-range order. In network polymers increasing amounts of long-range order are present. Diamond, the best example of a completely covalent network, is sometimes called a crystalline polymer.

X-ray diffraction patterns of polymers which show diffuse rings indicate short-range atomic order. Crystallinity is indicated when the pattern sharpens into spots. Such patterns are never as sharp

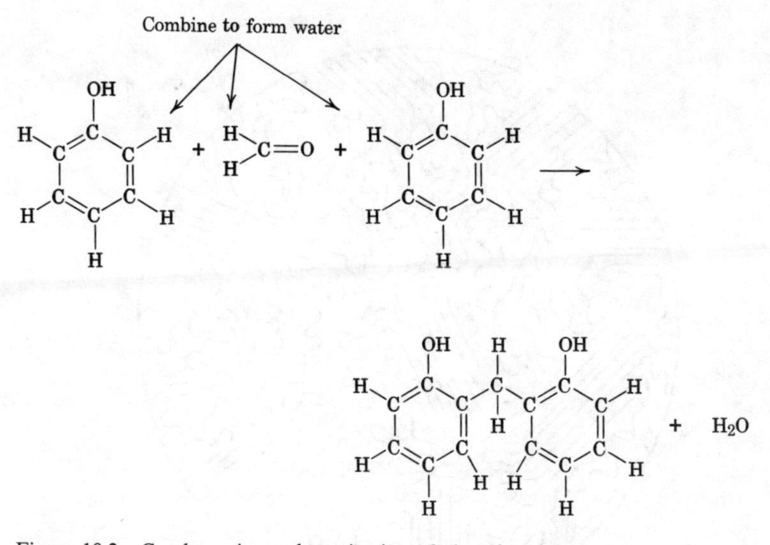


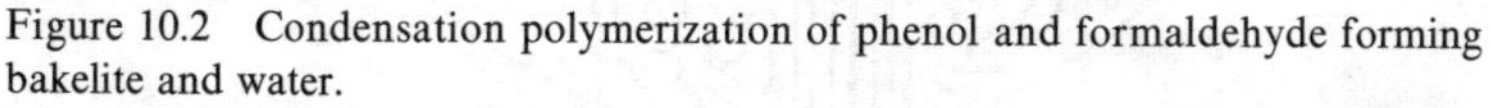

Figure 10.2 Condensation polymerization of phenol and formaldehyde forming bakelite and water.

as for a wholly crystalline material; furthermore, values for specific gravity calculated from the indicated atomic spacings are higher than the actual ones. Such polymers actually contain both crystalline and amorphous regions.

The relative amounts of crystalline and noncrystalline regions in polymers vary with the chemical composition, molecular configuration, and the processing. Early explanations for the crystalline-amorphous nature of polymers were based on the fact that X-ray measurements indicated crystallite sizes of the order of 100 Å, whereas the length of typical polymer molecules was many times this size. This suggested the so-called "fringe-micelle" mode, shown in Figure 10.3, in which an individual molecular chain may be packed with like molecular chains in one crystalline *micelle,* separated from similar micelles by amorphous regions. This model helps in visualizing the development of crystallinity in polymers during tensile elongation but is inadequate to explain the behavior of undeformed polymers.

Polyethylene can be dissolved in xylene in dilute solution and slowly cooled from above 120°C to form single crystals (see Fig-

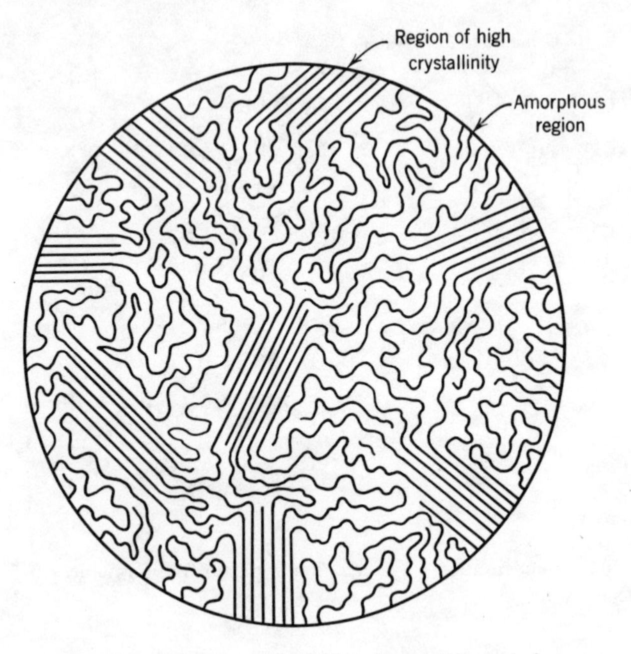


Figure 10.3 Fringed micelle model of crystallites in polymers.

ure 10.4*a*). The first crystals to form are thin diamond-shaped layers of very uniform thickness, usually of the order of 100 Å. X-ray examinations indicate that these layers are crystalline and that the polymer chains are perpendicular to the plane of the layers. Since the typical length of a chain is several times the thickness of the layer, the molecules must be folded back and forth through the crystalline region, as shown in Figure 10.4*b*. The surface of the layer must, then, be made up of "hairpin turns" of the molecular chains. Since these turns are chemically saturated, no primary bonds can be formed on the surface of the layers; adjacent layers, however, can be joined together by secondary (van der Waals) bonds.

When a large sample of a partially crystalline polymer is viewed by transmitted polarized light, a spherulitic structure, such as shown in Figure 10.5, is sometimes evident. This structure is typical of undrawn crystalline polymers of nylon and polyethylene. The spherulite appears to consist of crystalline ribbons which fan out radially from a center of spherical symmetry. A large number of such ribbons may be arranged to form a spherulite of visible dimensions (1 to 100 microns). As the size of the spherulite increases, polyethylene becomes more and more opaque. Because the spherulites are not crystallographically cubic, they are

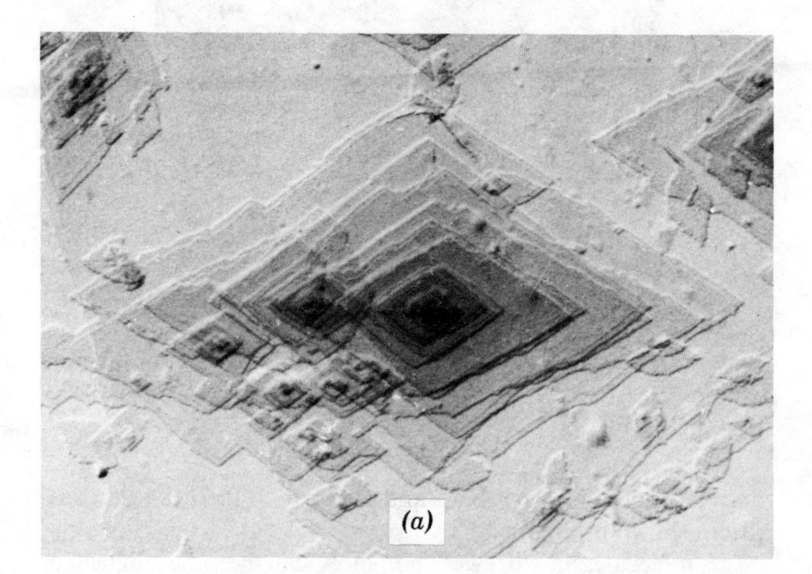


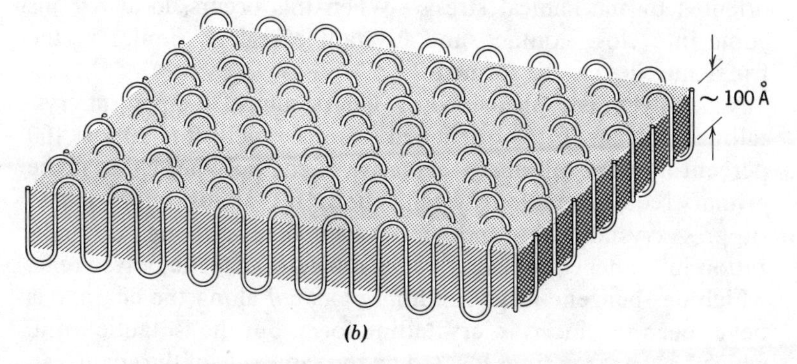


Figure 10.4 (*a*) Electron micrograph of a single crystal of polyethylene (from A. Keller, R. H. Doremus, B. W. Roberts, and D. Turnbull (eds.), *Growth and Perfection of Crystals,* John Wiley and Sons, New York, 1958, p. 499); (*b*) molecular structure of crystalline polyethylene.

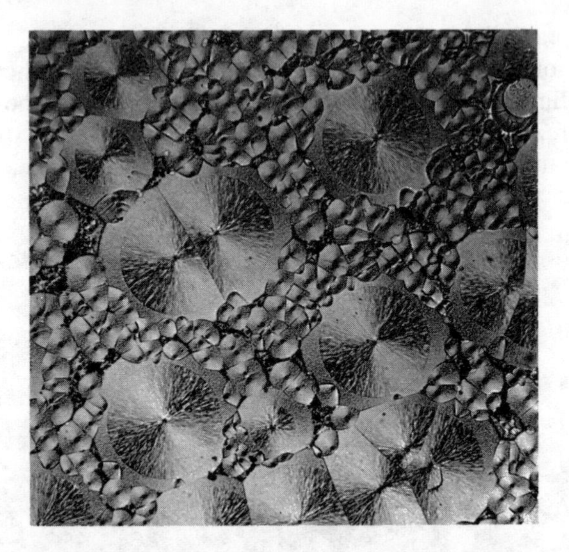

Figure 10.5 Spherulite structure of a crystalline polymer view between crossed polaroids (from F. P. Price in R. H. Doremus, B. W. Roberts, and D. Turnbull (eds.), *Growth and Perfection of Crystals,* John Wiley and Sons, New York, 1958, p. 466).

"active" under polarized light and give rise to the Maltese Cross pattern shown in Figure 10.5.

The molecules in "drawn" polymers can become aligned or oriented by mechanical stress. When this occurs, local regions come into close contact, and crystalline regions similar to the fringe micelles can be formed.

When nylon is cooled slowly, it can be almost 100 percent crystalline, whereas if it is quenched from the melt, it is almost 100 percent non-crystalline. Regularity of the polymer chain is the primary requirement for crystallization. It is almost impossible to suppress crystallization in linear polyethylene, although crystallization in branched polyethylene is difficult. Atactic polystyrene, which has benzene rings randomly located along the chain, has never been produced in crystalline form, but the isotactic form, which has benzene rings located on the same side of the chain, can be partially crystallized. The tendency for crystallization is also reduced by copolymerization and by the presence of unsaturated bonds or large side groups, which reduce the flexibility of the chain.

## 10.4 RESPONSE TO CHANGE IN TEMPERATURE

The drastic alteration which occurs in the behavior of polymers under stress over the range of temperatures from subzero to 200°C is of great practical significance and reveals a great deal about the role of structure in the response of a polymer to mechanical stress. Elastic modulus and specific volume are convenient properties by which to evaluate the behavior of polymers. It is important to note that elastic modulus is simply the coefficient of proportionality between stress and strain at an arbitrarily selected time of testing.

The variation of "elastic modulus," $E_r$, with temperature for polystyrene is shown in Figure 10.6. This curve is typical of polymers in general. The temperature ranges and levels, and even the

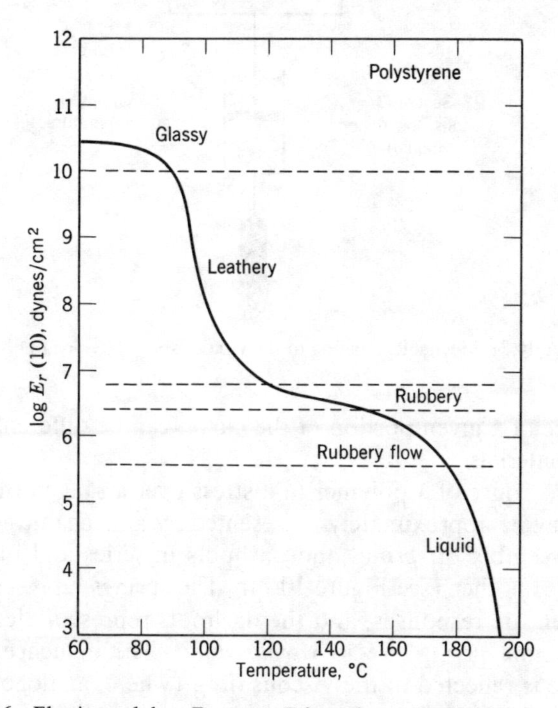


Figure 10.6 Elastic modulus, $E_r$, versus $T$ for polystyrene showing five regions of viscoelastic behavior.

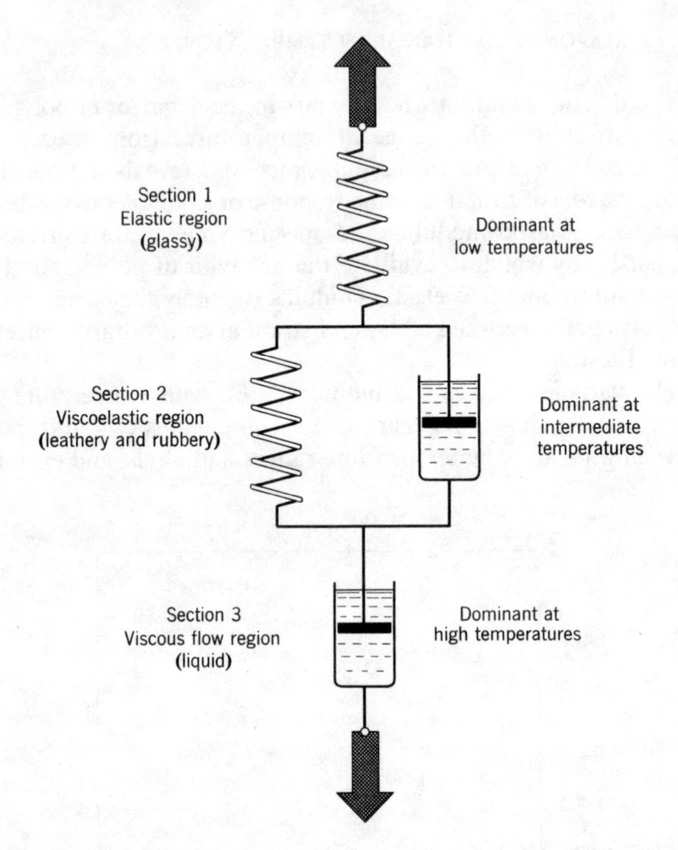


Figure 10.7 Mechanical analog to the viscoelastic behavior of polymers.

existence of a given portion of the curve, can be different for different materials.

The response of a polymer to a stress over a range of temperatures may be approximately represented by a mechanical analogy of an assembly of *springs* and dashpots in series and in parallel with one another (see Figure 10.7). The springs represent recoverable elastic responses, and the dashpots represent elements in the structure giving rise to *viscous drag.* The influence of temperature is reflected in the viscous drag (where, analogously, the viscosity of the oil in the dashpots decreases with increasing temperature).

In the *glassy* region (up to 80°C for polystyrene) a polymer is hard and brittle, and an applied stress produces a strain which is recoverable upon release of the stress. In this region, strain occurs by the stretching of bonds within and between molecular chains. The chains, which are "frozen" together, cannot flow past each other and may only be separated by fracture. The behavior of polymers in the glassy region resembles the deformation of the spring in Section 1 of Figure 10.7.

In the *leathery* region, where the modulus drops rapidly with temperature, reversible sliding becomes possible in short segments of the chain. Small sections move and then cause neighboring sections to move cooperatively. In this range a transition occurs between the elastic behavior of Section 1 and the viscoelastic behavior of Section 2 of the mechanical analog. The reversibility of the movements of the short-chain segments is expressed by the spring in Section 2 and the resistance to this movement, by the *dashpot.*

In the *rubbery* region of noncrystalline polymers, the viscoelastic behavior in Section 2 dominates the deformation. As the temperature increases above the glass transition temperature, the molecular segments slide reversibly past one another and tend to straighten out. In that class of materials called *elastomers* (which includes natural and synthetic rubbers) the "rubbery plateau" dominates the $E_r$ versus $T$ curve, and $E_r$ actually rises with increasing temperature. This increase, which is not typical of all polymers, is considered in a later section.

In the *rubbery flow* and liquid regions, permanent molecular sliding dominates the deformation process, as represented in Section 3 of the mechanical analog. As temperature increases, the viscosity decreases and the apparent modulus drops even more, to the extent that at high temperatures the polymer is essentially a liquid.

The rubbery region is not present in a polymer that is crystalline. The modulus in such a material decreases gradually with increasing temperature until the crystalline melting point is reached. This behavior is shown for crystalline isotactic polystyrene in Figure 10.8. The modulus of crystalline polystyrene is higher than that of the noncrystalline form at all temperatures above the glassy region. At about 225°C a sharp decrease in modulus indi-

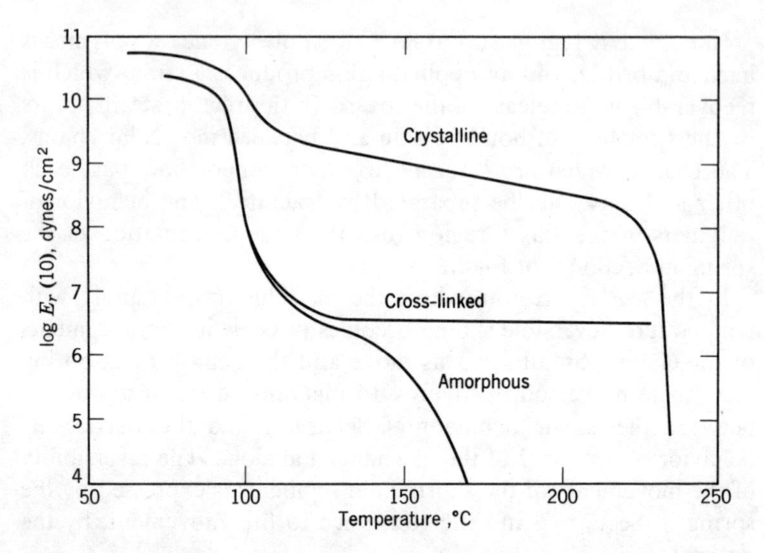


Figure 10.8 $E_r$ versus $T$ for crystalline isotactic polystyrene, cross-linked polystyrene, and amorphous polystyrene.

cates melting of the crystalline structure. In most polymers which contain both crystalline and amorphous regions, modulus versus temperature plots assume positions intermediate between the limits indicated in Figure 10.8. Rubbery behavior is generally observed when crystallinity is suppressed in molecular chains having bulky side groups or double bonds which prevent the free movement and alignment of chains.

The quantitative analysis of deformation in terms of a viscoelastic response is limited because real materials seldom behave as an ideal spring coupled with an ideal dashpot. However, their behavior can often be represented by a large number of springs and dashpots, as in Figure 10.7. If the springs are Hookean and the dashpots are Newtonian (these assumptions facilitate simple analysis, but both are inexact), the relaxation of force, $F$, as the deformation is held constant can be readily calculated from the Maxwell model:

$$F = F_0 e^{-(k/\eta)t} \tag{10.2}$$

From Equation 10.2, $F$ falls to $(1/e)F_0$ when $t = \eta/k$. This time,

$t = \eta/k$ is called the relaxation time of the system. Similarly, time-dependent deformation, $\delta$, which occurs under constant force, can be calculated easily for the Voigt model:

$$\delta = \delta_\infty[1 - e^{-(k/\eta)t}] \qquad (10.3)$$

where $\delta_\infty = F/k$ is the $\delta$ that would occur after infinite time. In this case, $\delta$ rises to within $1/e$ of its final value in a time $\eta/k$; $\eta/k$ is called the retardation time of the system.

It is possible to simulate the viscoelastic behavior of materials by various combinations of the above models with different values for $k$ and $\eta$. Usually the number required is too great for much utility, and the model concept is actually abandoned in favor of a more realistic idea: a distribution of relaxation times (or retardation times). The relaxation times, which can be measured by mechanical, acoustical, electrical, or electromagnetic means, may physically be associated with such responses as primary bond stretching, primary bond rotation, molecular chain straightening, chain segment displacement and entanglement, and side group motion.

## 10.5 THE GLASS TRANSITION TEMPERATURE

The transition from glassy to rubbery behavior in noncrystalline polymers appears in modulus-temperature measurements over a 10 to 20° temperature range. Figure 10.9 shows a plot of specific volume, $V$, versus temperature. A break occurs in the slope of this curve at the glass transition temperature, $T_g$, which marks the point above which molecular segments are free to move past one another, and below which they are confined.

The specific volume-temperature curve of crystalline polymers shows a discontinuity at the crystalline melting point. The glass transition is marked by a change in slope of the specific volume curve. At temperatures above the glass transition temperature, but below the melting point, crystalline polymers are rigid but not brittle. In this same range the noncrystalline regions in polymers exert their greatest influence in improving impact strength. The glass transition generally occurs at a temperature between one-half and two-thirds of the absolute melting point.

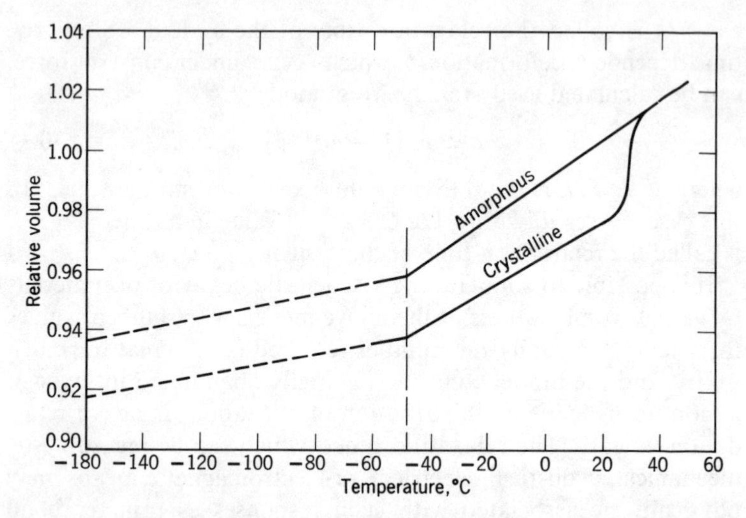


Figure 10.9 Relative volume as a function of temperature for amorphous and crystalline rubber (after N. Bekkedahl, *J. Res. Nat. Bur. Standards,* 13, 1934, p. 411).

## 10.6 ELASTOMERS AND RUBBER ELASTICITY

Most materials under stress exhibit an initial elastic region in which strain is proportional to stress; if the stress is released, the material returns to its original length. The amount of elastic strain does not ordinarily exceed 1 percent before some other mode of deformation sets in. Strain can be related to the elastic movement of atoms from their equilibrium positions, which in polymers can also involve the bending and rotation of C—C bonds.

Materials known as elastomers sustain reversible strains of several hundred percent. Figure 10.10 shows a typical stress-strain curve of an elastomer. Several significant features are characteristic of elastomeric behavior.

1. The material is soft, and its elastic modulus is low.
2. Very high strains are possible.
3. The strain is reversible.
4. The material is noncrystalline and is above its glass transition temperature.

While the "coiled molecule" picture explains the stress-strain curve of elastomers in a qualitative fashion, it does not provide a complete theory. There are several pieces of experimental evidence which it cannot embrace.

Early molecular theories attributed the behavior of elastomers to the postulate that the molecules were coiled in the unstressed state and were then capable of extensive elongation as they uncoiled under stress. This idea was further supported when it was observed that natural rubber (*cis*-polyisoprene) and gutta-percha (*trans*-polyisoprene) differed greatly in their properties because of their different molecular shape, shown in Figure 10.11. In gutta-percha, which is crystalline, hard and brittle, the molecular chain can readily be arrayed in linear form, whereas in natural rubber, because of the hydrogen and methyl groups on the same side of the nonrotating double bond, and the passage of the chain in and then out of the same side of the double bond, crystallization and chain straightening are restricted. Thus the chain of natural rubber is irregularly curved and tangled within the volume of the material, and large elongations are possible as the ends of the chain are pulled apart under stress.

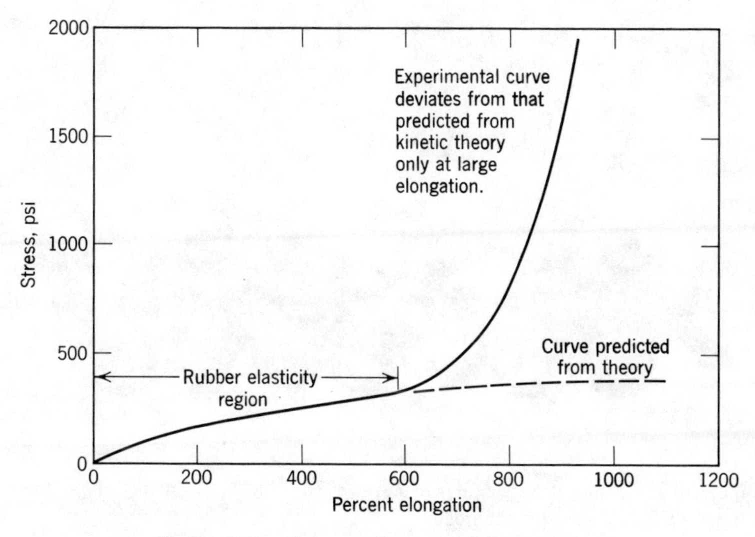


Figure 10.10 Stress-strain curve of an elastomer.

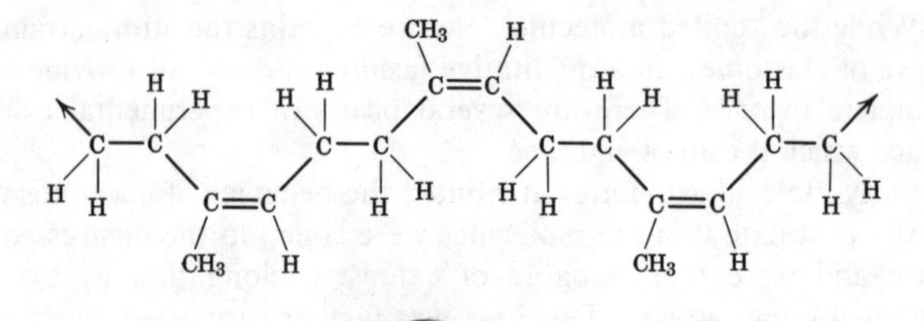

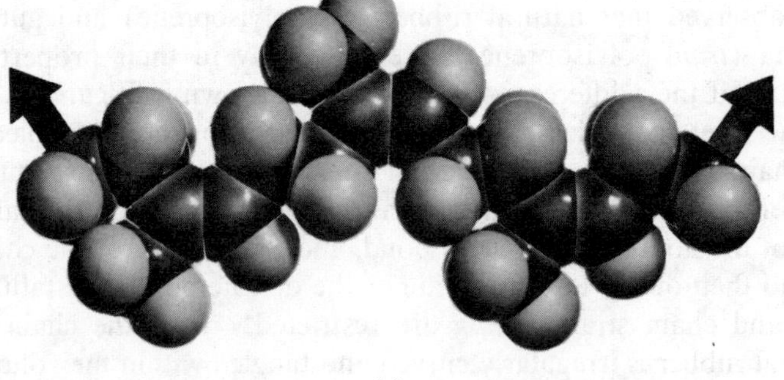

*(a)*

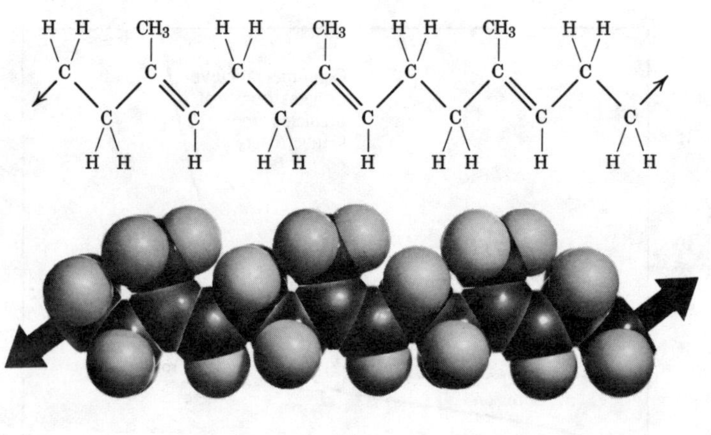

*(b)*

Figure 10.11 Molecular structures of (*a*) natural rubber (*cis*-polyisoprene) and (*b*) gutta-percha (*trans*-polyisoprene).

Three results contradict the behavior expected according to the coiled molecule model. First, the elastic modulus in the elastomeric range increases with increasing temperature; secondly, an elastomer becomes warmer rather than cooler when stretched rapidly; and, third, the coefficient of thermal expansion of an unstretched elastomer is positive while that of an elastomer stretched more than a few percent is negative.

The kinetic theory of rubber elasticity, proposed in 1935 by K. H. Meyer, helps to explain both the high recoverable deformation possible in elastomers and the thermal effects, for which the coiled spring model is inadequate. The deformation of an elastomer, in this theory, is analogous to the compression of an ideal gas. A combination of the first and second laws of thermodynamics applied to tensile strains leads to equation:

$$dU = T\,dS + F\,dL - p\,dV \qquad (10.4)$$

where $F$ is the tensile force and $L$ is the length of the sample. In rubbery materials Poisson's ratio is observed to be approximately $\frac{1}{2}$; thus tensile elongation does not cause a change in volume. If an ideal rubber is subjected to reversible isothermal extension ($dU = 0$), Equation 10.4 predicts the following relationship:

$$F = -T\left(\frac{\partial S}{\partial L}\right)_{T,V} \qquad (10.5)$$

Materials which exhibit rubbery behavior are those made of long-chain molecules in which neighboring chains are either mechanically entangled or cross-linked at various intervals along the chain. In the unstretched state, the portions of the chain between points of cross-linking or entanglement are randomly coiled and change rapidly from one coiled configuration to another (if all possible configurations are consistent with the fixed distance between cross-links). When the chain is extended, the distance between cross-links increases, thus reducing the number of possible coiled configurations from $\Omega_0$ to $\Omega$. This leads to a decrease in entropy, as stated by the relation:

$$S_0 - S = k \ln \frac{\Omega_0}{\Omega} \qquad (10.6)$$

This change in entropy can be stated as

$$S_0 - S = \frac{1}{2} N_0 k \left[ \left( \frac{L}{L_0} \right)^2 + 2\left( \frac{L_0}{L} \right) - 3 \right] \tag{10.7}$$

where $N_0$ is the number of chains between cross-links in a sample whose original length and cross-sectional area are $L_0$ and $A_0$ and whose length and cross-sectional area in the stretched state are $L$ and $A$. Differentiating Equation 10.7 with respect to length, and substituting this value in Equation 10.5, gives a tensile force which obeys the relationship

$$F = \frac{N_0 kT}{L_0} \left[ \frac{L}{L_0} - \left( \frac{L_0}{L} \right)^2 \right] \tag{10.8}$$

Dividing the tensile force by the cross-sectional area gives the true stress

$$\begin{aligned} \sigma_T &= \frac{F}{A} = \frac{N_0 kt}{AL} \left( \frac{L}{L_0} \right) \left[ \frac{L}{L_0} - \left( \frac{L_0}{L} \right)^2 \right] \\ \sigma_T &= nkT \left[ \left( \frac{L}{L_0} \right)^2 - \frac{L_0}{L} \right] \end{aligned} \tag{10.9}$$

where $n$ is the number of chain segments between cross-links per unit volume. The *isothermal modulus of elasticity,* defined as $E = L(\delta\sigma_T/\delta L)_T$, can be obtained by differentiation of Equation 10.9:

$$E = nkT \left[ 2\left( \frac{L}{L_0} \right)^2 + \frac{L_0}{L} \right] \tag{10.10}$$

This theory of rubber elasticity agrees fairly well with experimental findings except at very large elongations. At this point the molecular chains begin to align themselves, and further deformation results from the stretching of primary bonds within the molecular chains instead of by the straightening of originally coiled chain segments.

Vulcanization, which is the formation of sulfur cross-links between adjacent chains at positions where double bonds originally existed, extends the temperature range of the rubbery region and also increases the glass transition temperature. Lightly vulcanized soft rubber contains a sulfur cross-link for every several hundred chain units. Hard rubber contains 25 to 30 percent sulfur by weight and the available double bonds are saturated; softening does not occur until temperatures near 100°C and the material can no longer be considered an elastomer.

## 10.7 ORIENTATION EFFECTS

The response to stress of both elastomers and ordinary polymers changes after very large deformations. The first several hundred percent elongation in the stress-strain curve in Figure 10.11 was attributed to rubber elasticity. Just before fracture, the stress-strain curve deviates from the theoretical curve and rises steeply. This sudden increase in stress required to produce more strain is due to the exhaustion of the extensibility of the individual molecules and the development of crystallinity from the alignment of the chains in the direction of stress. X-ray diffraction and optical activity (birefringence) studies have shown that oriented crystalline regions form at high strains.

The molecular chains in nonelastomeric polymers also become aligned under strain. This is commonly revealed by the occurrence of a yield point in the stress-strain diagram and is known as "drawing" or "cold-draw." Figure 10.12 shows a typical stress-

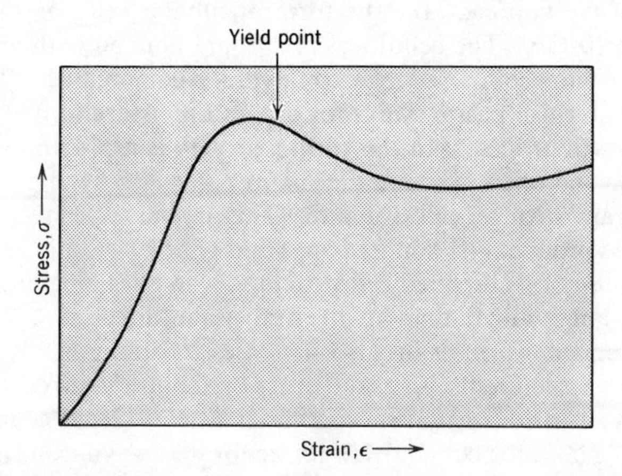


Figure 10.12 Stress-strain curve for undrawn nylon showing yield point due to the unfolding of molecular chains.

strain for an undrawn polymer. The yielding is thought to be due to the instability in deformation produced when the folded crystals of the undrawn material unfold progressively. The material is crystalline in the drawn condition and the molecular chains are oriented in the direction of drawing. As a result, the longitudinal strength of the drawn fiber is many times that of the undrawn material.

## 10.8 WOOD

Wood, among natural polymers, is a material whose mechanical properties are dominated by the fact that its molecular chains are both strongly oriented and crystalline. The molecular structure of wood is based on the cellulose chain represented in Figure 10.13*a*. Typical natural cellulose products have three to four thousand $C_6H_{10}O_5$ units in the chain and, because of the polar nature of the hydroxyl groups on the molecule, are highly crystalline. The assembly of molecules decomposes before it softens at elevated temperatures. If mechanical processing is required, cellulose can be treated chemically to reduce the molecular weight and neutralize the polar influence of the hydroxyl groups.

In wood, cellulose crystals, which comprise only 50 to 60 percent of the volume, take the form of tubular cells, as shown in Figure 10.13*b*. The cellulose crystals are bonded with an amorphous lignin composed of carbohydrate compounds. The cell cavities furnish passageways for sap and moisture to flow during the growth process. In the spring, growth is rapid and the cells are thin and open, while in the summer slower growth produces a denser and stronger cell structure. This annual cycle in the growth of wood produces its familiar rings and grain.

Because of its directed cell structure, wood has greater strength in the longitudinal than in the transverse direction. A highly dense and uniform grain also increases its strength. While the cellulose structure itself has a density of about 1.5 gm/cc, the bulk density of various wood species varies from 0.12 gm/cc for balsa to 0.74 gm/cc for oak and 1.3 gm/cc for lignum vitae. The hardness and strength of these woods vary similarly. The high affinity of wood for moisture is a significant factor in controlling its

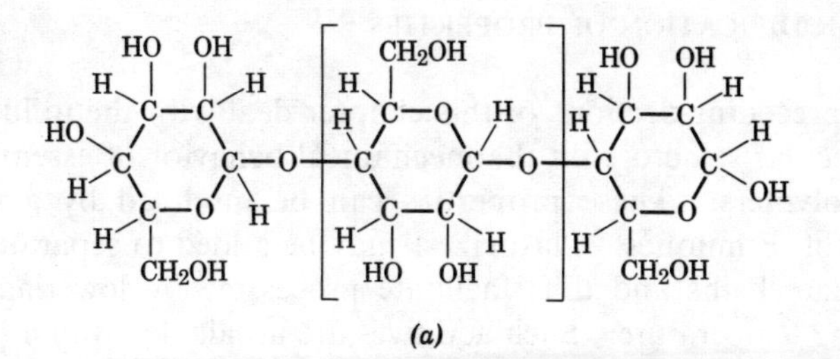


(a)

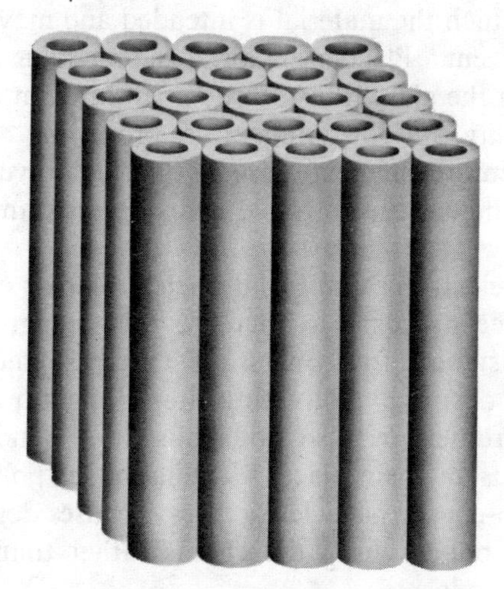

(b)

Figure 10.13 (*a*) Molecular chain structure of cellulose; (*b*) cellular structure of wood.

dimensional stability. Wooden articles warp in the process of drying, as water is more readily evaporated from sides perpendicular to the grain than from those parallel to the grain. Many techniques have been employed to optimize the directional properties and minimize the moisture absorption of wood. These include lamination, cross-lamination as in plywood, plastic impregnation and high-temperature compression with binders.

## 10.9 MODIFICATION OF PROPERTIES

The preceding sections of this chapter deal with the influence of molecular structure on the mechanical behavior of essentially pure polymers. These properties can be modified by a wide variety of techniques. Plasticizers may be added to separate the molecular chains and thus facilitate processing by lowering the softening temperature. Such additives are usually low vapor pressure materials which are miscible with the polymer. The use of polypropylene glycol in polyvinyl chloride and polyvinyl acetate is typical of this technique. The amounts of plasticizer added depend on the use for which the material is intended and may be as much as 35 weight percent. Plasticizers lower the modulus and strength but, by lowering the glass transition temperature, improve impact resistance, ductility, and low-temperature processing.

Fillers and reinforcements are usually inert additives. They are made to lower the cost, to improve abrasion resistance, strength, and color.

High voltage electron, X-ray, and nuclear reactor radiation has been used in recent years to increase the strength and high-temperature resistance of polymers. The increase in cross-linking in polyethylene occurs by activating sites along adjacent chains. Neutron bombardment in this case breaks C—H bonds and forms new C—C bonds in their place. Irradiation of polytetrafluoroethylene by high-energy particles or X-rays causes degradation by breaking C—C bonds along the chains rather than promoting stronger C—F bonds.

## 10.10 DESIGN CONSIDERATIONS

Polymeric materials have unique enough properties that an ever-increasing number of applications have resulted. Rubbers, both natural and synthetic, in tires and tubing, and wood in furniture are familiar examples. Many engineering applications exist where the strength to weight ratio is responsible for the substitution of polymers for metals. Where poor thermal or electrical conductivity is desirable, they also find widespread use. In certain wear-resistant applications such as bushings, molybdenum disulfide filled polyamides have been found superior to various metal bushings from the standpoints of both wear and coefficient of friction. There are specific characteristics of resins which always must be considered before specifying their use. Their low thermal conductivity and high coefficient of expansion require that parts subjected to high loads and speeds be designed with great care. Failure of such parts can often be attributed to incomplete dissipation of frictional heat. The intelligent designer consequently specifies thin walls on sleeve bearings, for example. In all applications he likewise specifies adequate clearance to permit thermal expansion. He furthermore must consider that most polymers deteriorate in excess heat, light, or in contact with various chemical agents. In all cases, he must know their rate of deterioration in order to recommend their use intelligently in any new specific application.

## DEFINITIONS

*Addition Polymers.* Polymers formed into long chain molecules by the chemical reaction of one or more types of monomer units, each of which has at least one double bond prior to polymerization.

*Condensation Polymers.* Polymers formed by the chemical reaction of at least two monomer units with the production of a by-product of low molecular weight.

*Copolymer.* An addition polymer of at least two monomers.

*Degree of Polymerization.* An expression of the length of the average molecular chain in a polymer; usually stated in terms of the number of repeating units in a chain.

*Branching.* The inclusion of large atoms, radicals, or molecular chains on the side of a parent molecular chain.

*Cross-Linking.* The formation of covalently bonded segments between adjacent long molecular chains.

*Bifunctional.* Having two sites for chemical reaction per monomer.

*Thermoplastic Polymers.* Polymers capable of plastic deformation at elevated temperatures without breakdown.

*Thermosetting Polymers.* Polymers which, once produced, cannot be softened and worked at elevated temperatures without chemical or structural decomposition.

*Spherulites.* Crystalline regions in an undrawn polymer visible under polarized light.

*Glass Transition Temperature.* The temperature at which a sudden change in slope of the specific volume versus temperature curve occurs. It very nearly approximates the temperature below which a polymer fails in a brittle manner and above which it behaves as a leathery or rubbery solid.

*Elastomer.* A polymer characterized by very high reversible elongation, a negative coefficient of expansion, and positive temperature coefficient of modulus.

*Plasticizers, Fillers, Reinforcements.* Materials deliberately added to polymers to enhance their properties.

## BIBLIOGRAPHY

### Introductory Reading

Alfrey, T., Jr., "Molecular Structure and Mechanical Behavior of High Polymers," in J. E. Goldman, *Science of Engineering Materials,* John Wiley and Sons, New York, 1954, p. 469.

Jastrzebski, Z. D., *Engineering Materials,* John Wiley and Sons, New York, Chapters 2, 4, 12.

Orowan, E., "Polymers," in *Mechanical Behavior of Materials,* by F. A. McClintock and A. S. Argon, MIT, 1962.

Tobolsky, A. V., "Mechanical Properties of Polymers," *Scientific American* **197** (September, 1957).

Van Vlack, L. H., *Elements of Materials Science,* Addison Wesley, Reading, Mass., 1964, Chapter 7.

### Supplementary Reading

Alfrey, T., Jr., *Mechanical Behavior of Polymers,* Interscience Publishers, New York, 1948.

Billmeyer, F. W., *Textbook of Polymer Science,* Interscience Publishers, New York, 1962.

Flory, P. J., *Principles of Polymer Chemistry,* Cornell University Press, 1953.

Nielsen, L. E., *Mechanical Properties of Polymers,* Rheinhold, New York, 1962.

Tobolsky, A. V., *Properties and Structures of Polymers,* John Wiley and Sons, New York, 1960.

Treloar L. A. C., *The Physics of Rubber Elasticity,* Oxford University Press, London, 1949.

## PROBLEMS

10.1 Summarize briefly how each of the following modifies the strength of a polymer:

(a) Increased degree of polymerization.
(b) Increased branching.
(c) Increased cross-linking.
(d) Increased crystallinity.

10.2 What feature is necessary in a monomer for addition polymerization to be possible? Is this same feature a necessity for condensation polymerization?

10.3 Describe the difference between thermoplastic and thermosetting polymers in terms of:

(a) Applied stress.
(b) Increased temperature.
(c) Atomic structure.

10.4 On the basis of the viscoelastic model of a polymer, show why the following occur:

(a) Elastic moduli measured at different rates have different values.
(b) Elastic moduli measured at different temperatures have vastly different values.
(c) Constant stress applied for a very long time always causes continuing deformation.

10.5 A sample of polyethylene sinks in mineral oil at room temperature. If the oil temperature is raised above 120°C, the polyethylene becomes clear and floats. Explain.

10.6 The average molecular weight of a polyvinyl chloride has been determined to be 9500. What is the degree of polymerization (number of repeating units in the average molecule)?

10.7 Natural rubber is a polymer of isoprene: $H_2C{=}\underset{CH_3}{\underset{|}{C}}{-}\underset{H}{\underset{|}{C}}{=}CH_2$.

(a) Show how it polymerizes.
(b) What feature of the polymer chain permits vulcanization?

10.8 When polyisoprene (natural rubber) in the soft, partially vulcanized form is exposed to air for long periods it gradually becomes hard and cracks easily. Suggest a reason for this behavior.

10.9 The oxidation resistance of butyl rubber is superior to that of natural rubber. Butyl rubber is a vulcanized polymer of isobutylene and one or two percent isoprene. Suggest a reason for its oxidation resistance.

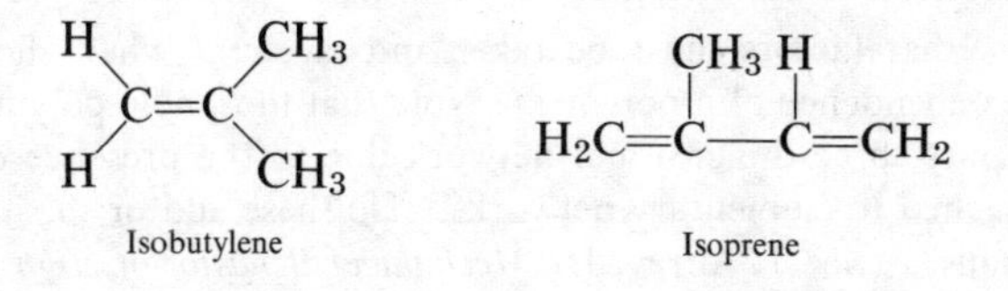


10.10 Unvulcanized natural rubber can be stretched to just below its breaking point and, when released, returns to its original length. If it is quenched in cold water before releasing, it retains its stretched length. Explain this behavior.

10.11 In a particular rubbery material, if the number of moles of chain segments between cross-links is $10^{-4}$ moles/cc, what is the initial elastic modulus? What is the elastic modulus after 300 percent extension in length?

10.12 Will increased cross-linking lower or increase the elastic modulus of an elastomer? Why?

10.13 Explain how advantageous mechanical properties are attained by the cross-lamination process in the manufacture of plywood.

10.14 Prepare a short paragraph which compares the nature of crystallinity in polymers with that in metals and ceramics.

10.15 Write a brief essay on the useful and harmful effects of radiation on polymers. (See *Effect of Radiation on Materials,* edited by J. J. Hammond et al., Rheinhold Publishing Co., New York, 1958.)

10.16 The deformation of single crystals of polyethylene indicates that while single crystals deform, the following processes occur: (1) slip, (2) twinning, (3) phase changes. Which is rate controlling? (See *J. Appl. Phys.,* **35** (1964) 1599.)

10.17 (a) Plot strain at constant stress against time for different polymeric material including (1) low molecular weight, (2) high molecular weight, (3) slightly vulcanized (under-cured), (4) lightly vulcanized, and (5) highly vulcanized materials.

(b) To what are the initial similarity and subsequent divergence of these curves ascribed to?

10.18 (a) When an ideal gas is compressed isothermally, how do its energy and its entropy change?

(b) Ideal gas molecules are compressed in a container. What does thermal agitation do to the system?

(c) What is the energy of rubber in the stretched state as compared to its energy in the unstretched state?

(d) Why is the stretching of rubber thermodynamically analogous to the compression of an ideal gas?

10.19 What factors must be taken into account when showing the structure dependence of a polymer? Note that the whole polymeric solid is not a pure three-dimensional network due to the presence of "linear tails" attached to elementary networks. Do these add or substract from the modulus? (See T. Alfrey, Jr., *Mechanical Behavior of High Polymers,* Interscience Publishers, 1948.)

10.20 (a) Draw a stress versus time curve for a thermoplastic polymer.

(b) Draw a strain versus time curve for a thermoplastic polymer, indicating total strain.

(c) On the same plot draw a viscous strain versus time curve.

(d) Now plot the strain versus time for a high elasticity polymer.

10.21 (a) Draw a load-elongation curve for rubber showing hysteresis.

(b) What characteristics shown in this curve account for the value of rubber as a shock absorber?

# Index

## SELECTED PHYSICAL CONSTANTS

| | | |
|---|---|---|
| Speed of light | $c$ | $3.00 \times 10^8$ meters/sec |
| Avogadro's number | $N_0$ | $6.02 \times 10^{23}$ molecules/mole |
| Gas constant | $R$ | 8.32 joules/(mole)(°K) = 1.98 cal/(mole)(°K) |
| Planck's constant | $h$ | $6.63 \times 10^{-34}$ joule-sec = $6.63 \times 10^{-27}$ erg-sec |
| Boltzmann's constant | $k$ | $1.38 \times 10^{-23}$ joule/°K |
| Permeability constant | $\mu_0$ | $1.26 \times 10^{-6}$ henry/meter |
| Permittivity constant | $\epsilon_0$ | $8.85 \times 10^{-12}$ farad/meter |
| Electron charge | $e$ | $1.60 \times 10^{-19}$ coul = $4.8 \times 10^{-10}$ statcoulomb |
| Electron rest mass | $m_0$ | $9.11 \times 10^{-31}$ kg |
| Electron charge-to-mass ratio | $e/m_0$ | $1.76 \times 10^{11}$ coul/kg |
| Mass-energy relation | $c^2 (= E/m)$ | 931 Mev/amu = $8.99 \times 10^{16}$ joules/kg |
| Magnetic moment of the electron | $\mu_e (\equiv \mathfrak{M}_B)$ | $9.29 \times 10^{-24}$ joule-m$^2$/weber |

## SELECTED CONVERSION FACTORS

1 radian = 57.3° = 0.159 rev
1 kilogram = 2.21 lb (mass)
1 pound (mass) = 0.454 kg
1 atomic mass unit = $1.66 \times 10^{-27}$ kg
1 meter = 39.4 in. = 3.28 ft; 1 inch = 2.54 cm
1 mile = 5280 ft = 1.61 km
1 angstrom unit = $10^{-10}$ meter = $1 \times 10^{-4}$ microns
1 millimicron = $10^{-9}$ meter
1 atmosphere = 29.9 in. Hg = 76.0 cm Hg = $1.01 \times 10^5$ Nt/meter$^2$
1 Btu = 778 ft-lb = 252 cal = 1055 joules
1 calorie = 4.19 joules; 1 joule = 0.239 cal = $2.78 \times 10^{-7}$ kw-hr
1 electron volt = $1.60 \times 10^{-19}$ joule
1 horsepower = 550 ft-lb/sec = 746 watts